The Basics of Climatology

Steven Richardson

Edward A

First Published 1986
by Edward Arnold (Publishers) Ltd
41 Bedford Square
London WC1B 3DQ

Edward Arnold (Australia) Pty Ltd
80 Waverley Road, Caulfield East
Victoria 3145, Australia

British Cataloguing in Publication Data
Richardson, Steven
The basics of climatology.
1. Climatology
I. Title
551.6 QC981

ISBN 0-7131-7475-7

Set in 10/11 point Times by
Anneset, Weston-super-Mare, Avon
Printed in Great Britain by
Butler & Tanner Ltd, Frome, Somerset and
bound by W H Ware & Sons Ltd, Clevedon, Avon

Acknowledgements

The Publishers wish to thank the following for permission to reproduce copyright material:

Barnaby's Picture Library (Max and Kit Hunn) p.27
The National Remote Sensing Centre p. 28
The Meteorlogical Office p. 33, 40

The author would like to acknowledge Dave Bowyer, Bill Ellis, Roger Hardy and especially thanks Pat Thomas.

Dedication
To Pat and Tom

Contents

Preface

The study of climate is an area which is often overlooked at A-level, and yet it is climate which is partly responsible for the formation of soils thereby controlling the distribution of vegetation and the development of landscape. Climate is therefore a major factor contributing to human activities such as agriculture, recreation, transport, construction and planning.

Our understanding of climate has been enhanced since the Second World War, especially with the advent of high-flying aircraft and satellites, the latter also having helped to provide more accurate information for weather forecasters. However, there are still many areas where theory and understanding are less than perfect.

The knowledge of climate demanded by most syllabuses has changed rapidly: for example, ten years ago A-level students were not expected to know about jetstreams, yet today they may be expected to discuss them at interviews. However, our understanding of jetstreams is still changing, explanations are continually being abandoned or rejected. Add to this the problems of people changing the climate by destruction of rainforests, or the burning of fossil fuels, and creating new climates in urban areas and you have a most complex and dynamic area of study.

Whilst it is not possible in a book of this size to cover all these topics in great depth, it is possible to convey a good understanding of the mechanisms by which climates are produced.

The book begins with an examination of the sun's energy, followed by a study of the ways in which solar energy receipts vary over the earth's surface. It is argued that variations in the amounts of energy received are responsible for the production of patterns of circulation in the atmosphere and oceans, which, in turn, distribute excess energy to areas of the earth which have an energy deficiency. The final three chapters concentrate on the ways in which seasonal energy availability, distance from the oceans, and the migration of large-scale convection cells produce different climates.

This book demands an active approach to learning and understanding. There are many questions integrated into the text and at the end of Chapter 6 there are further questions designed to encourage independent research into climatology and associated areas.

1 Solar Energy and the Atmosphere

The sun

The source of most of the energy responsible for heating the earth's surface, the oceans and the atmosphere is the sun. The sun is a small star, and it produces energy as a by-product of the combination of light elements—such as hydrogen. This process is called nuclear fusion. This combination of atoms and the associated release of energy causes a loss of mass in the sun of 4.5 million tonnes per second. This is insignificant when compared with the total mass of the sun.

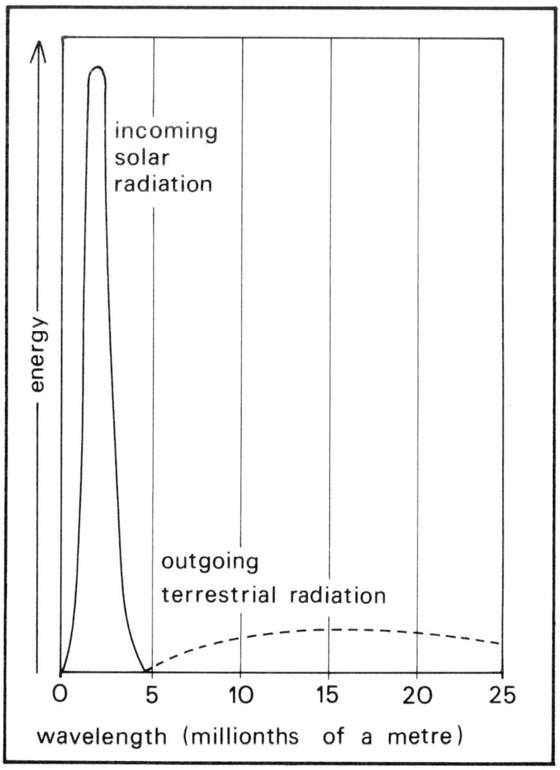

Figure 1 Solar and terrestrial radiation

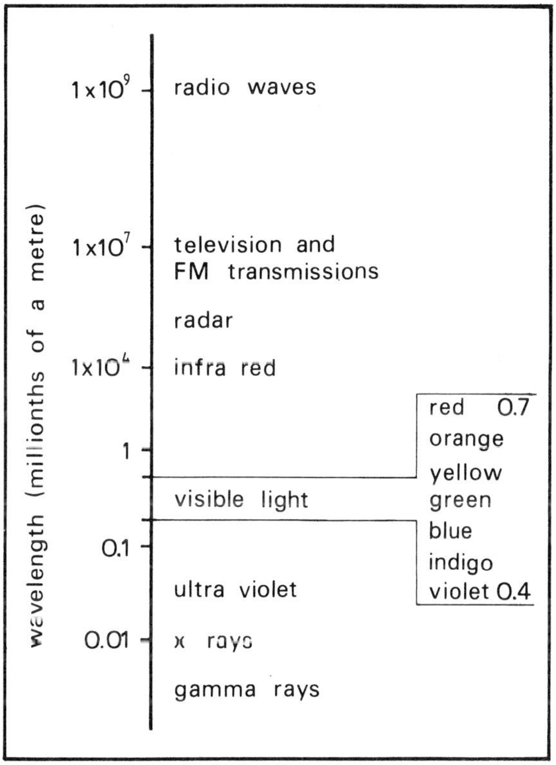

Figure 2 The electromagnetic spectrum

The energy produced by the sun is called insolation. The surface temperature of the sun is thought to be about 5727°C. This is very hot and at this temperature the sun gives out many forms of energy, but most of this insolation has a very short wavelength (see Figure 1). It is obvious that very hot objects give out energy; it may be both seen and felt. However, an object which is at −40°C is able to radiate energy. This may seem improbable, but such an object is still 233°C above absolute zero, the temperature at which objects do not radiate energy. If the sun, a very hot object, radiates energy of short wavelength and high

intensity, then as objects of lower temperatures radiate energy, the radiation they produce is of longer wavelength and lower intensity.
Examine Figures 1 and 2 carefully.

---

1 What are the wavelengths of visible light?
2 In which ways do insolation and the energy radiated by the earth differ?
3 Name two types of radiation produced by the sun, but not produced by the earth.

The atmosphere

The atmosphere is thought to be about 1000km deep, but the lowest 32km contain about 99% of the atmosphere. Most of the atmosphere is therefore very thin — 1% concentrated in a layer 968km deep! The composition of the atmosphere varies with altitude. On average slightly less than 80% of the atmosphere is nitrogen, just less than 20% is oxygen, with small amounts of water vapour and negligible amounts of argon, ozone, krypton, neon and helium. The proportions of oxygen and nitrogen remain constant throughout, but most water vapour and carbon dioxide are found below 10km, while most ozone is found above 25km.

Figure 3 shows a cross-section of the atmosphere, and the ways in which temperature and pressure vary with altitude. The names of the various layers in the atmosphere end with the

Figure 3 The structure of the atmosphere

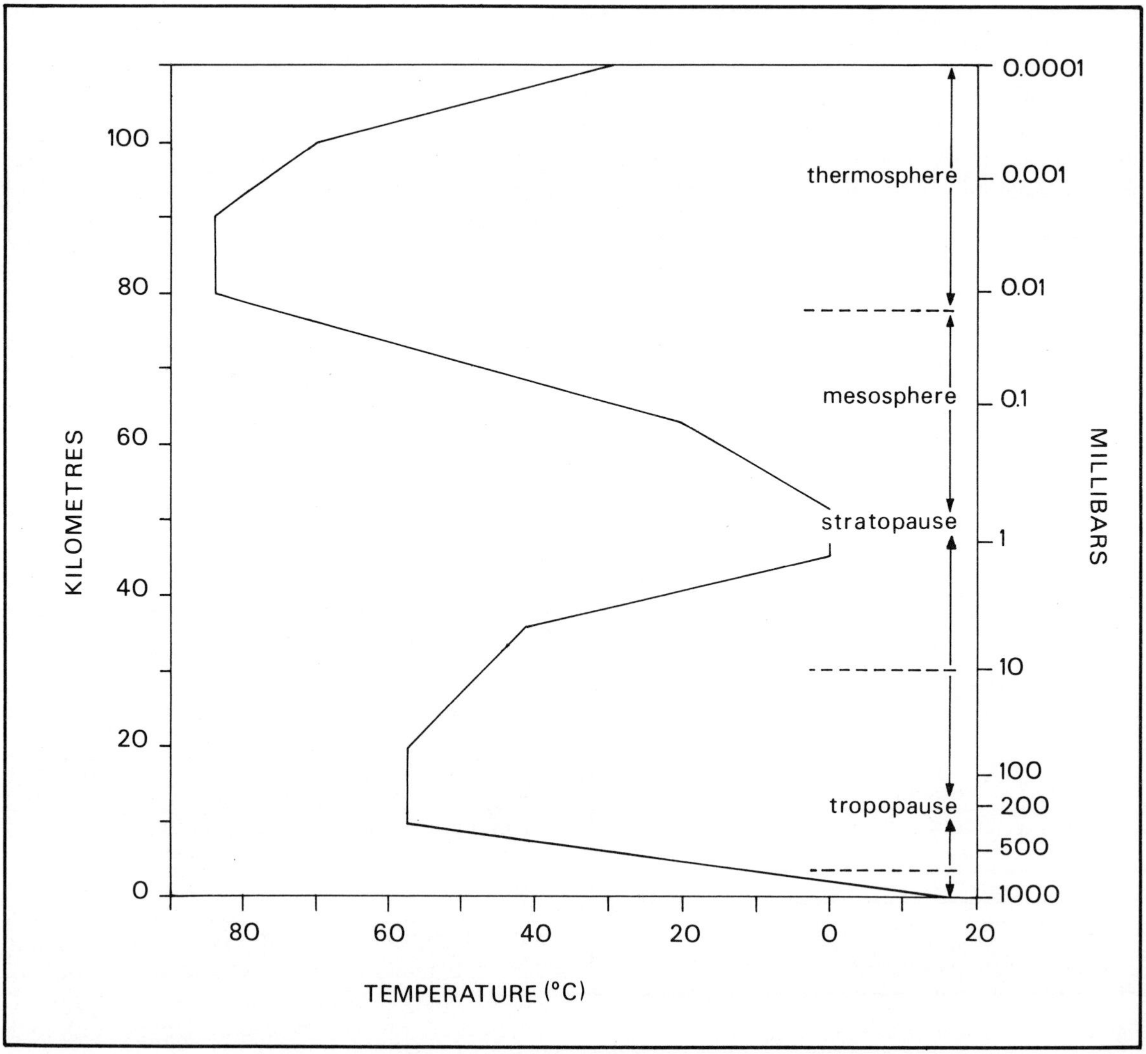

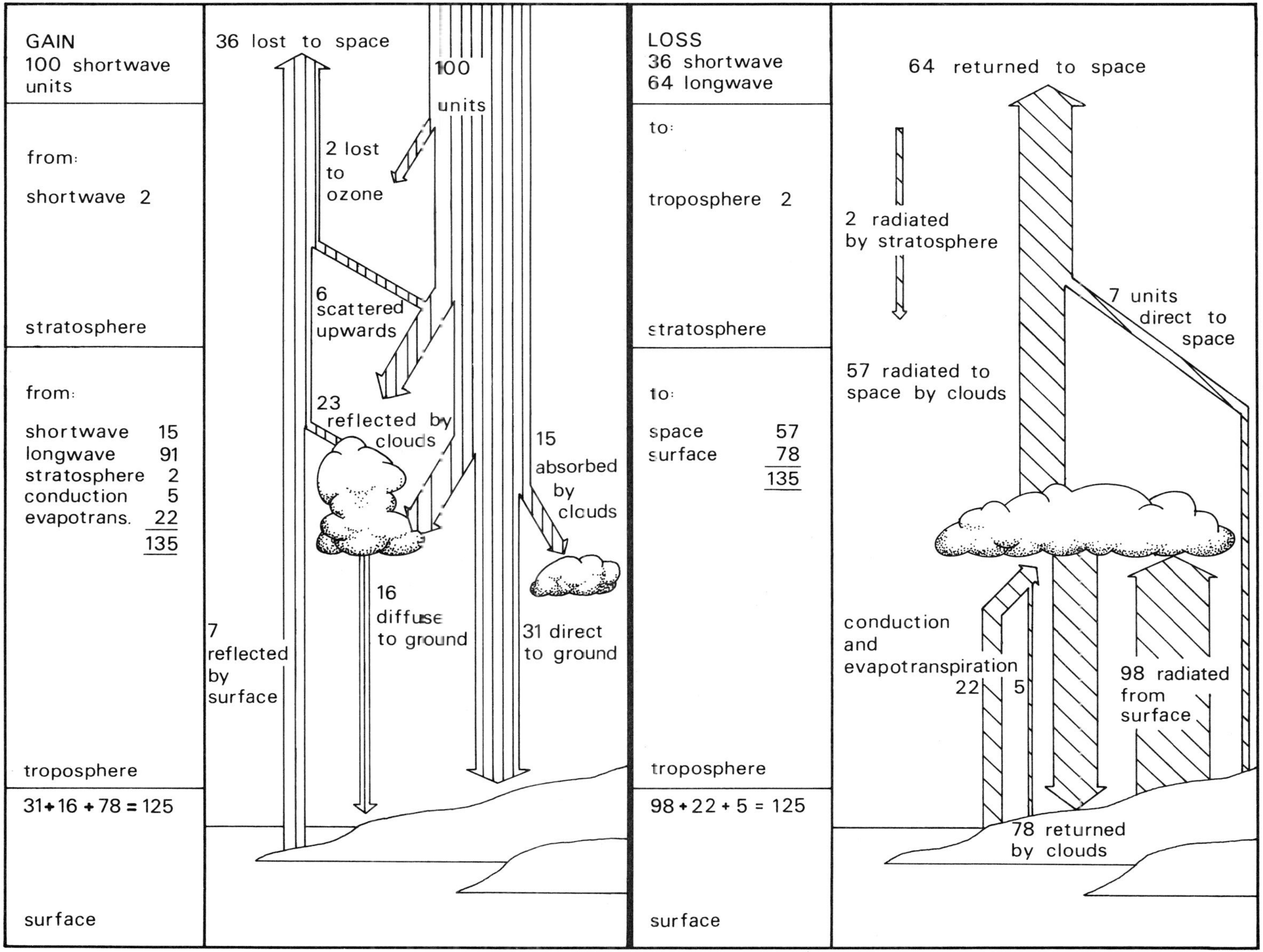

Figure 4 Gains and losses of energy

suffix "sphere" and the boundaries in "pause".

---

4 Make a copy of Figure 3, filling in the names of the missing spheres and boundaries.
5 What is the temperature and pressure at the boundary between the two lowest layers?
6 Describe the relationship between temperature and altitude. Can you suggest any possible reasons for this relationship?
7 To what do you attribute heating below 10km?

Some insolation is absorbed, scattered and reflected. Only certain types of insolation are absorbed, e.g. ultra violet, while all types are scattered. Some infra-red radiation is absorbed by water vapour and carbon dioxide. Much of the absorbed insolation is re-emitted in the form of radiation — usually of the longwave variety (see Figure 1).

Generally the wavelength of most insolation is so short that most atmospheric particles do not absorb it. This means that not all the insolation at the top of the atmosphere will reach the earth's surface. Figure 4 shows what happens to insolation as it passes through the atmosphere. Examine Figure 4 carefully, noting in particular the balance between incoming insolation and outgoing radiation and insolation.

---

8 Why is it necessary for energy gains to equal losses?
9 What percentage of insolation is reflected by clouds?
10 What percentage of insolation reaches the surface of the earth?
11 What percentage of insolation is absorbed before it reaches the earth's surface?
12 Of the radiation returned to space, what percentage is returned by clouds?
13 How important are conduction and evapotranspiration?

The amount of energy reflected at the surface depends upon the nature of the surface. Snow and ice reflect a great deal of insolation, up to 95% in the case of fresh snow. Sandy soil reflects between 25 and 45%, while coniferous forests reflect only 10–15%. The name given to this reflectivity is **albedo.**

---

14 Make a list of albedos of different surfaces.

The earth, once it has absorbed either insolation or re-emitted longwave radiation, acts as a giant radiator, constantly absorbing more energy to replace that which it has lost to space. The oceans and land surfaces are heated, and ocean and air currents distribute this heat to cooler latitudes. This heat is eventually radiated to space, but clouds are able to trap and return it to the earth. Look at Figure 4 again.

---

15 What percentage of radiation is returned to the earth by clouds?

If this energy was not returned, the average temperature of the earth would be −20°C. This energy is used by plants in evapotranspiration, and in helping to maintain the hydrological cycle. This energy also helps to produce convection currents in the atmosphere and oceans.

2 The Earth–Sun Relationship

The Earth–Sun Relationship

Chapter 1 dealt largely in averages, but the earth does not have an average climate. There are many different climates, produced in part by the variations in the amounts of insolation received at different places on the earth's surface. Receipts of insolation vary over the earth because the relationship between the earth and the sun is not a constant one.

The average distance of the earth from the sun is 149·5 million km. However, the orbit of the earth is not a circular one, and the earth is nearer to the sun in January — at 147 million km — than it is in July — at 152 million km. This means that the insolation reaching the earth is 7% less intense in July than it is in January.

The amount of energy received per unit area over the earth's sunlit side varies with latitude, and if the passages of three "sun beams" are followed through to the earth's surface, as in Figure 5, the reasons become apparent. Examine Figure 5 carefully.

Figure 5 Latitude and intensity of insolation

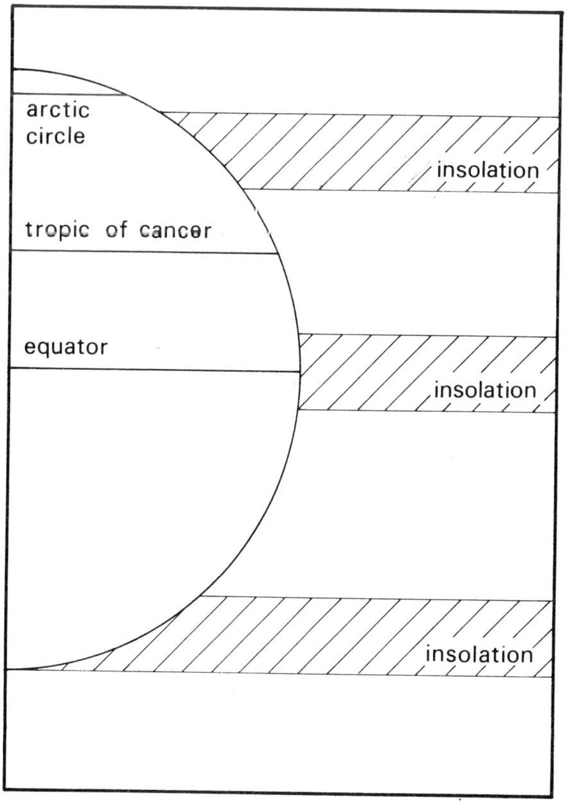

1 At which latitudes is insolation least intense? Explain your answer clearly.
2 Complete the following statement: "Low latitudes receive the most insolation because ...."

In addition to latitudinal variations in receipts of solar energy, the movement of the earth's axis also influences the amount of energy received. This axis of rotation — through the poles — is not at right angles to the plane of the earth's orbit. Figure 6 shows the tilting of the earth's axis at specific times of the year.

3 From the diagram, measure the maximum tilt of the earth's axis away from the vertical.
4 Between what latitudes does insolation reach the earth's surface at an angle of 90° for a part of the year?
5 Draw two diagrams to show the tilt of the earth's axis on September 22nd and June 21st.
6 Assess the importance of the tilting of the earth's axis and latitude on the receipt of energy.

The tilting of the axis affects the amount of insolation received during the year at any one latitude. Examine Figures 6 and 7.

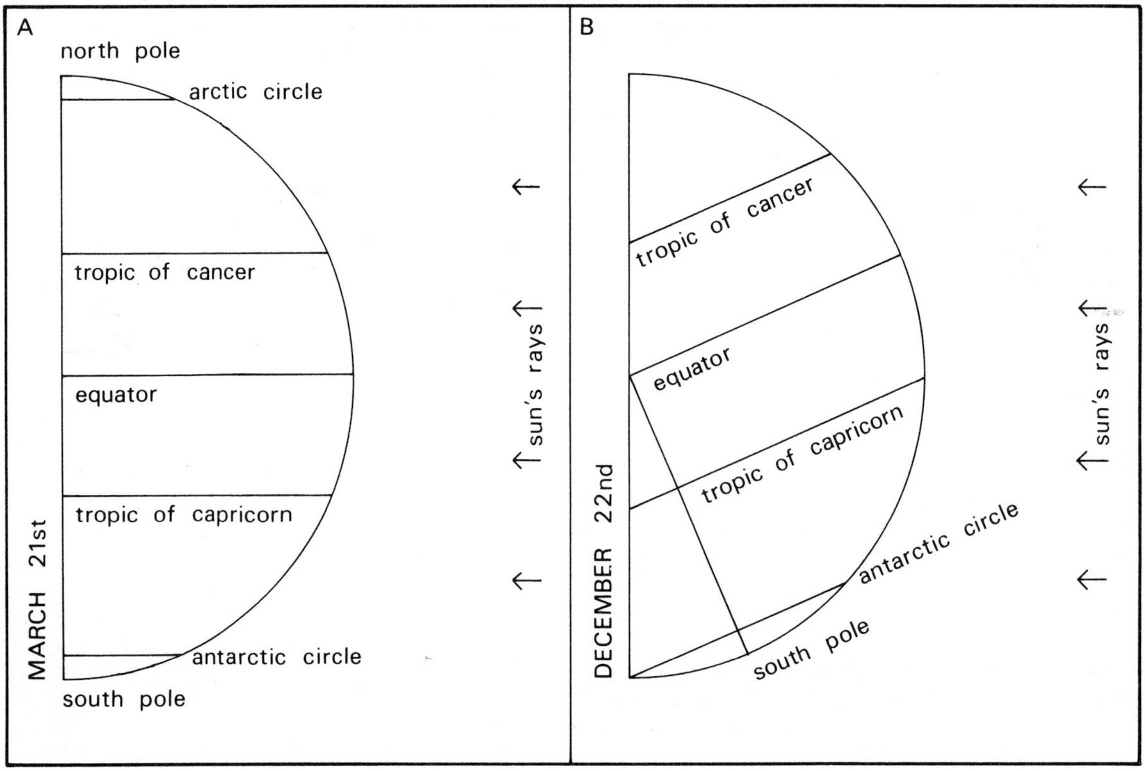

Figure 6 The earth at equinox and December solstice

7 Explain why at 60°N there are about six hours of daylight on December days and about eighteen hours in the days of late June.

8 What is the approximate duration of darkness at 80°S?

9 When the sun is overhead at 23½°S, what is the duration of daylight at 45°N?

10 Why, despite as much as six months continuous daylight, do the polar regions not receive as much solar energy as the tropics?

11 If the earth's axis of rotation was always perpendicular to the plane of its orbit, how long would day and night be at every place on the earth's surface?

12 Copy and complete Table 1, indicating the appropriate solstices and equinoxes for both hemispheres.

13 When it is the first day of summer in the southern hemisphere, what is the maximum and minimum duration of daylight in Canada?

Table 1

	LATITUDE OF OVERHEAD SUN	LENGTH OF DAYLIGHT 40 N	LENGTH OF DAYLIGHT 60S	EQUINOX OR SOLSTICE
MARCH 21st				
JUNE 21st				
SEPTEMBER 22nd				
DECEMBER 22nd				

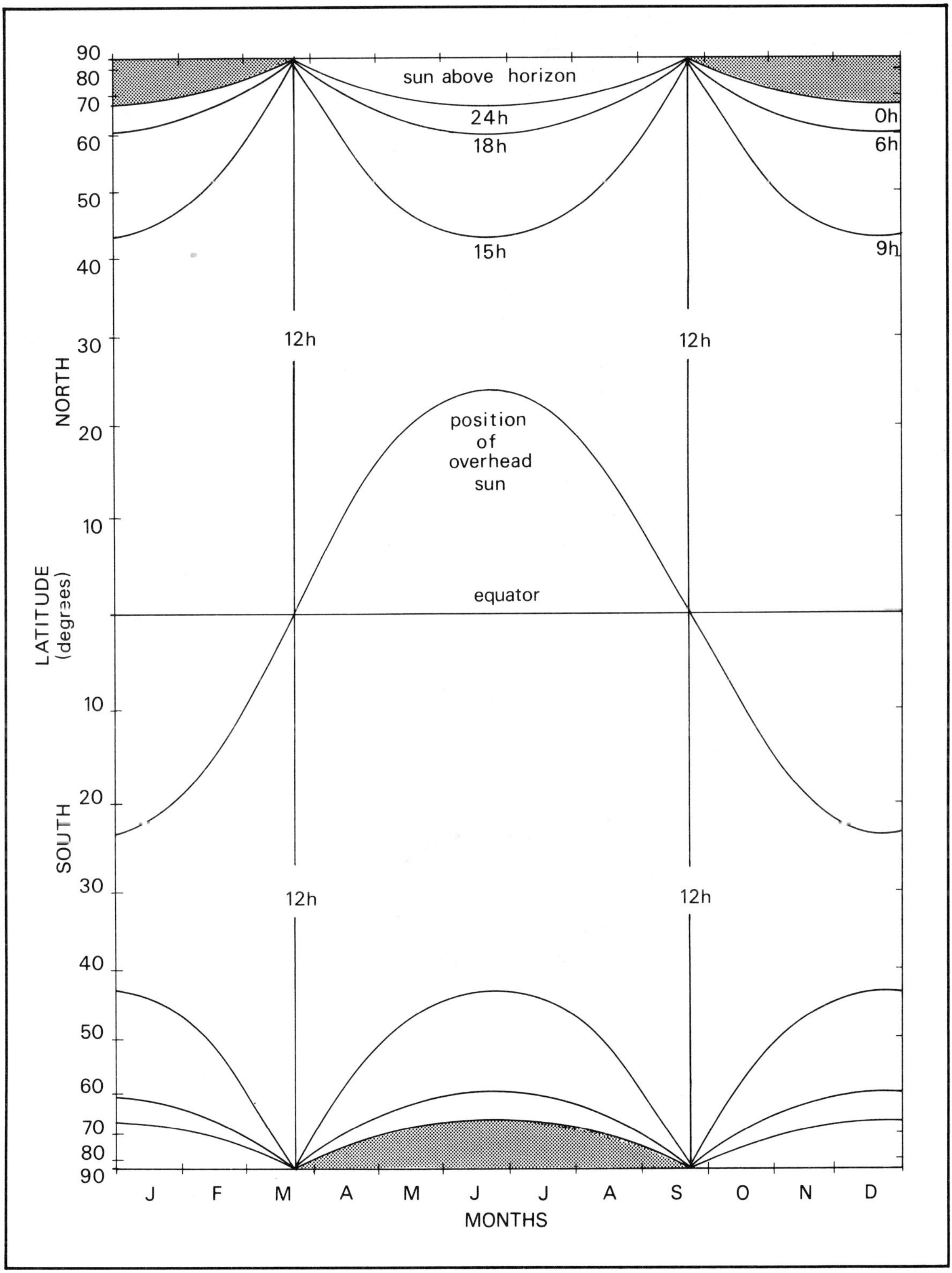

Figure 7 The variation of daylight hours with latitude

In summary, the tilting of the earth's axis and the curvature of the earth's surface bring about marked seasonal variations in heating at middle and high latitudes. At 90° North and South weak heating takes place for as long as six months in the year, but cooling may also take place for at least this length of time. In the middle latitudes there is cooling in the winter (low-sun period), but even so the sun may provide weak heating in these short summer days. The sun is high in the sky in mid-latitude summers providing reasonably intense heating. In the tropics there is never a marked cooling period as the sun is always high in the sky and lengths of day and night vary little. Returning to Chapter 1 and the discussion of solar energy, it is apparent that the earth radiates heat at all times because its temperature is greater than absolute zero. Some latitudes absorb more energy than others, so it would be expected that at some latitudes energy is lost, while at others it is gained. Figure 8 shows the balance between incoming shortwave radiation and outgoing longwave radiation.

14 Describe the pattern of receipts and losses of energy between the equator and 90°.

15 At approximately which parallel of latitude are

Figure 8 The variation of energy gains and losses with latitude

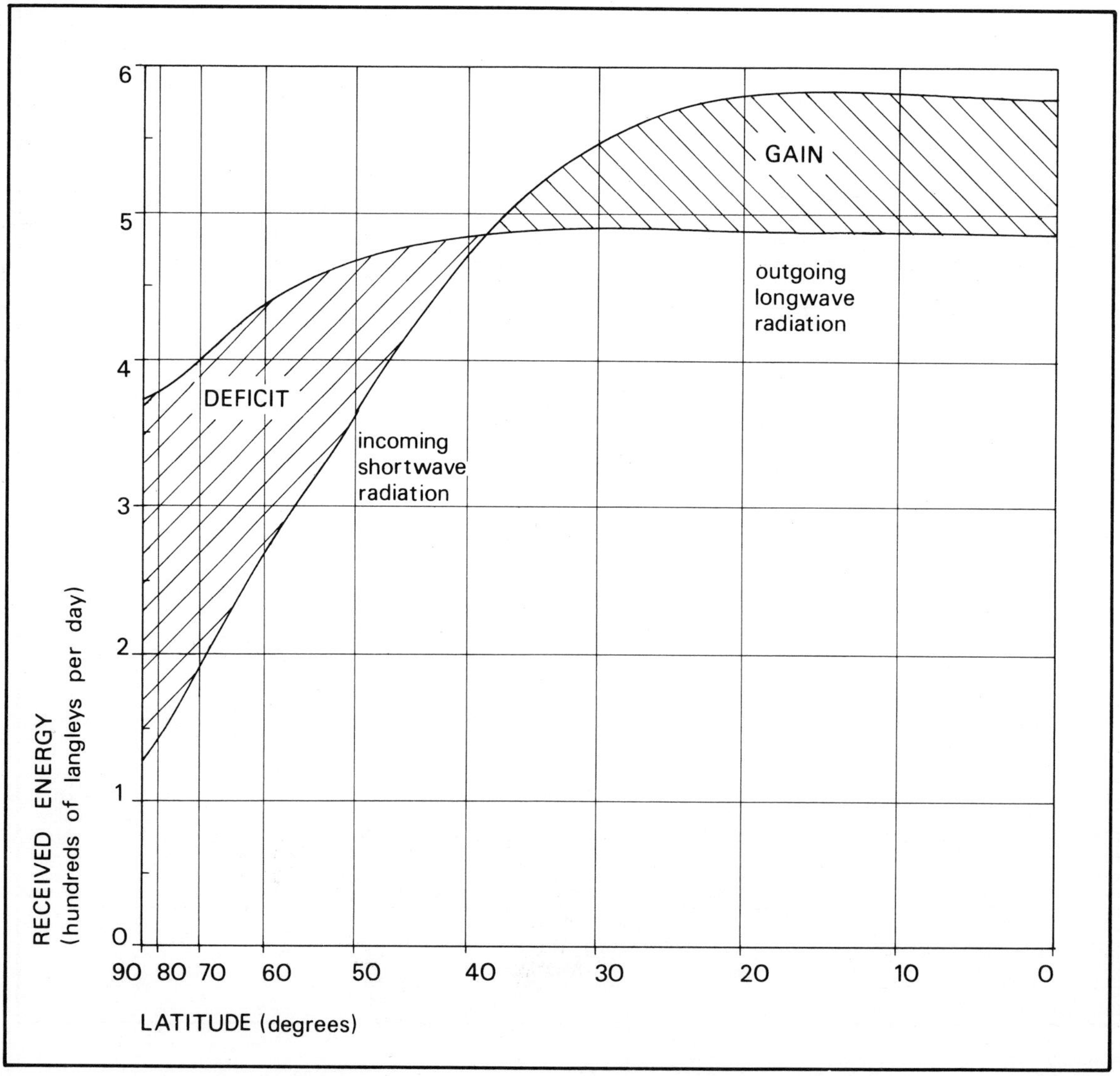

receipts and losses of energy balanced?

16 Which latitudes have a surplus of energy?

17 Make a copy of Figure 8, and on it indicate the direction of transfer of the surplus of energy.

The low latitudes receive more energy than they lose by radiation. They therefore have an excess of energy which, if it were not dissipated would result in the equatorial regions being 14°C warmer than they actually are, and the polar regions 25°C colder.

The transfer of surplus energy to regions which are deficient is the subject of the next chapter, for it is the energy differences which are partly responsible for the circulation of fluids. At present we must consider briefly the uses to which energy is put locally.

Some energy arriving at the earth's surface is used directly by plants in transpiration or in the melting of ice and the evaporation of surface water. The sun's energy and the radiated longwave energy heat the oceans and land surfaces and evaporate water which is then "transported" to other regions. At the same time, the energy used to evaporate the water is transported, for when the water vapour condenses, the energy is given back to the atmosphere. Energy is removed in the process of evaporation and returned to the atmosphere in the process of condensation. Examine Figures 9 and 10.

---

18 Where and in what form is most fresh water stored?

19 Approximately what percentage of fresh water is stored in the atmosphere at any one moment?

20 Describe the hydrological cycle.

Figure 9 The earth's water

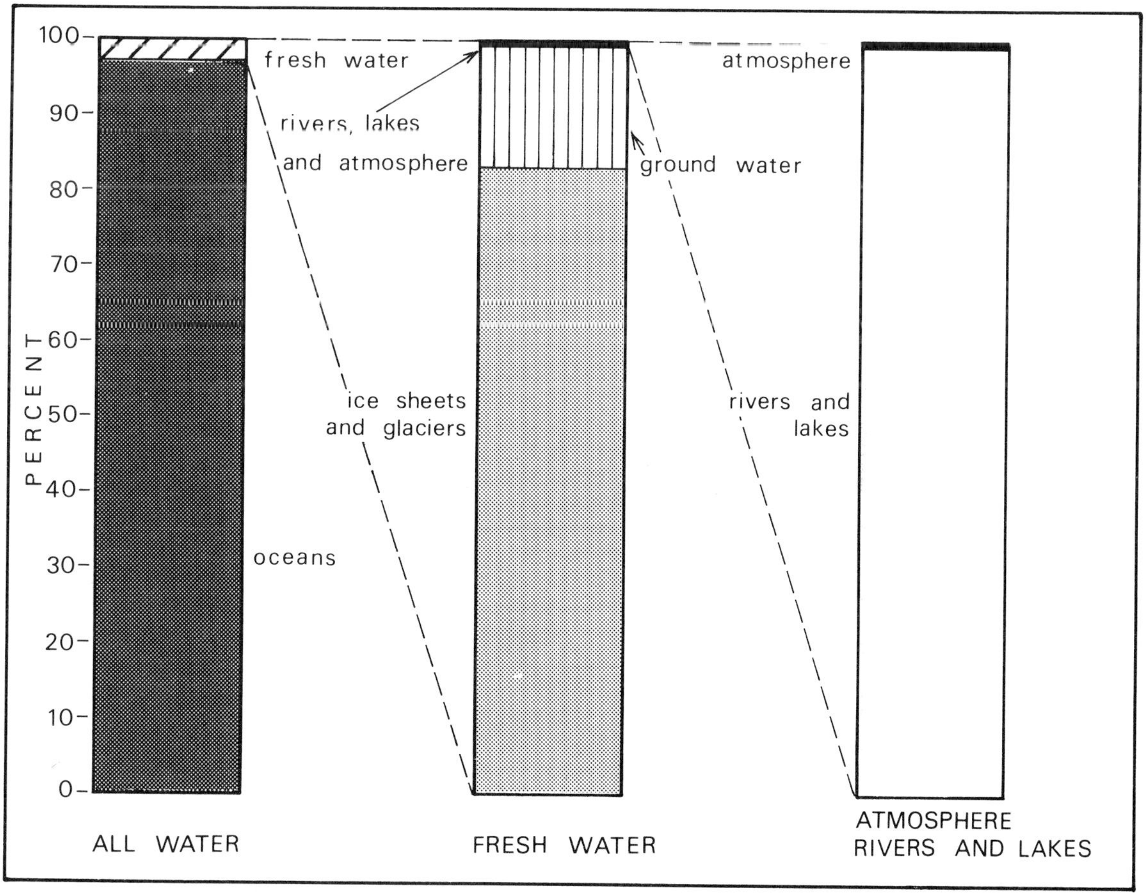

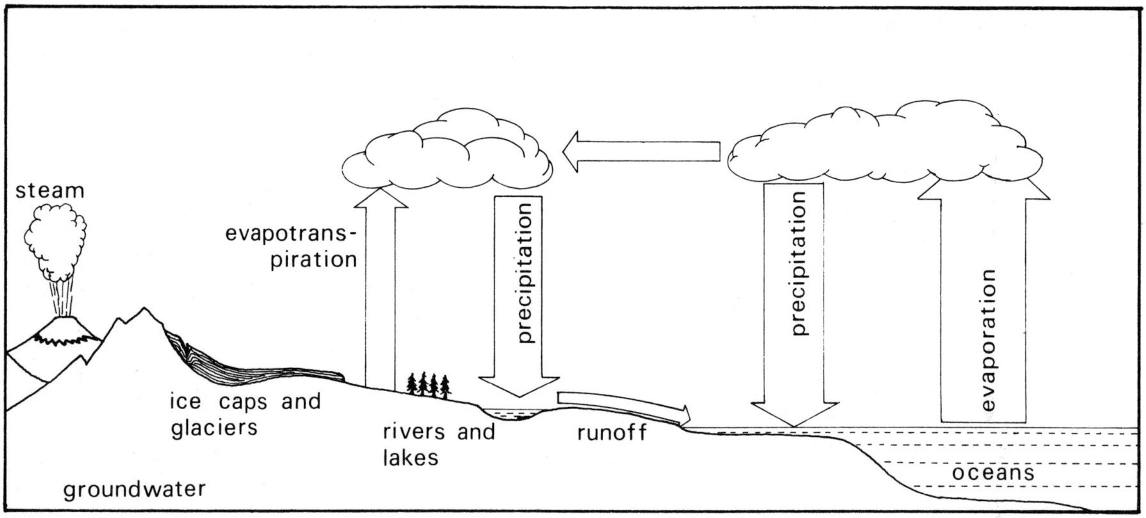

Figure 10 The hydrological cycle

3 Atmospheric and Oceanic Circulation

Chapter 2 described the ways in which heating varies both seasonally and latitudinally over the earth's surface. These variations, to a large extent, provide the force which drives atmospheric and oceanic currents. Nevertheless, it is not immediately obvious how this energy imbalance resolves with other forces to produce permanent or semi-permanent circulation patterns. In this chapter, therefore, the effects of energy imbalance, of rotation and of pressure difference on atmospheric and oceanic circulation, will be examined.

Energy imbalance and atmospheric circulation

A further examination of Figures 7 and 8 will show that energy receipts and therefore heating over the surface of the earth are very uneven. These inequalities had long been recognized, but it was not until the advent of high flying aircraft that their full implications were understood.

Until the middle of the 18th Century scientists thought that air and water must take heat from the tropics and transport it directly towards the poles, because if this were not so, the tropics would become increasingly hot and the poles increasingly cold. The first simple explanations showed air moving from the equator to the poles in the upper atmosphere.

---

1 Make a copy of Figure 11 and complete the pattern of circulation. Explain why this pattern of circulation may and may not be possible.

2 Using an atlas, name any two wind belts in the northern and southern hemispheres which indicate that this conception of circulation is inaccurate.

Sailors had long understood that this pattern of circulation (shown in your completed diagram) did not exist. The pattern of flow at the surface had long been known, but it was not fully appreciated that the surface flow was part of a three dimensional flow containing many "open circuits" of movement. The problem was that until almost half way through the 20th Century it was impossible to obtain information about what was happening in the upper atmosphere.

Looking at Figure 11 it is possible to see why this pattern of circulation could exist, if the air only descends to the surface of the earth at the poles as it cools. In reality, however, cooling takes place well before the air reaches the poles.

---

3 Make a copy of Figure 12.
  (i) Explain why the air rises and descends in the pattern shown, and
  (ii) With the help of an atlas, indicate with the words HIGH and LOW, the pressure at latitudes 0° and 30°, and name the winds which blow between these two pressure belts.

Figure 12 shows a **convection cell** caused by the intense heating in low latitudes. This pattern of circulation was first suggested by George Hadley — an English scientist — in 1735, and is called a **Hadley Cell**.

Air is a poor conductor of heat, and is unable to take up the sun's energy immediately. Therefore there is a delay between the sun being at its highest and temperatures being at their hottest.
Look back to Figure 7 (p.11).

---

4 By how many degrees of latitude does the zone of most intense heating vary during the year?

5 Are the Hadley Cells likely to follow precisely the position of the overhead sun? Explain your answer carefully.

6 Draw three diagrams to show the approximate positions of the Hadley Cells in both hemispheres at the solstices and equinoxes.

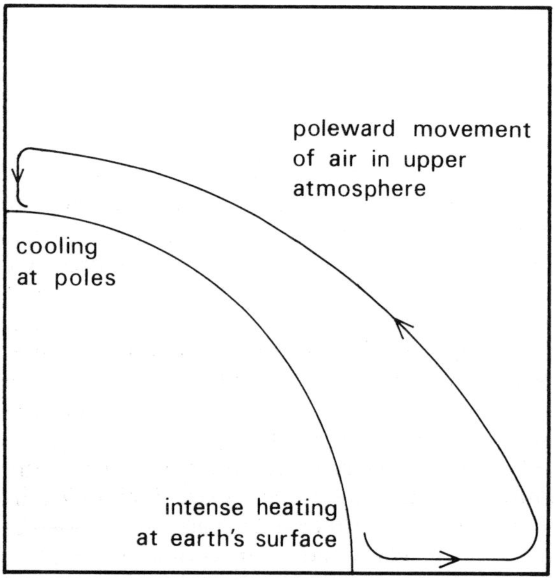

Figure 11 A simple model of atmospheric circulation

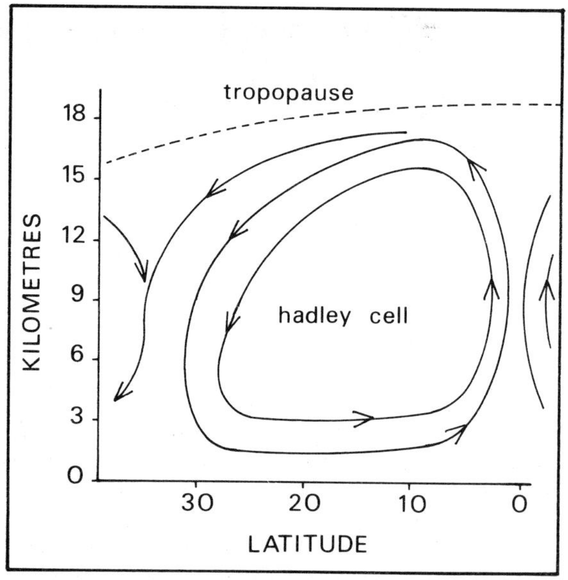

Figure 12 The Hadley Cell

7 Why is there a greater seasonal variation in the latitude of the thermal equator over land than over the ocean?

The surface convergence of the air moving towards the equator, i.e. the Trade Winds, is called the Inter Tropical Convergence Zone. The ITCZ moves north or south with the seasonal movement of the area of most intense heating — often called the thermal equator. This area is shown in Figure 13.
Examine Figure 13.

The circulation of air in the tropics and subtropics is dominated by the Hadley Cell, and it is this cell which is also responsible for the distribution of energy into higher latitudes. The poleward limits of the Hadley Cells mark the starting points of the Ferrel Cells.

The **Ferrel Cell** distributes heat from the tropics and subtropics to the mid and high latitudes. The air in both the Hadley and Ferrel Cells is tropical. The air in the Ferrel Cell is warm and is initially

Figure 13 The seasonal movement of the ITCZ

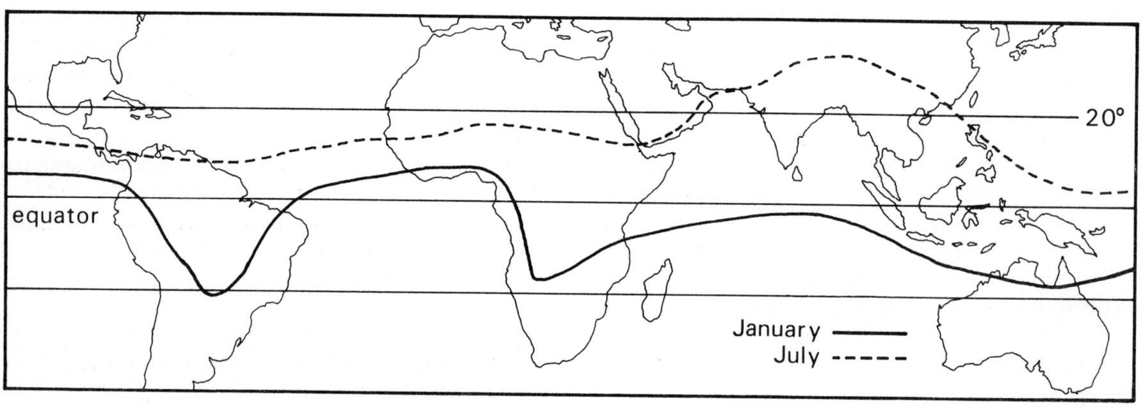

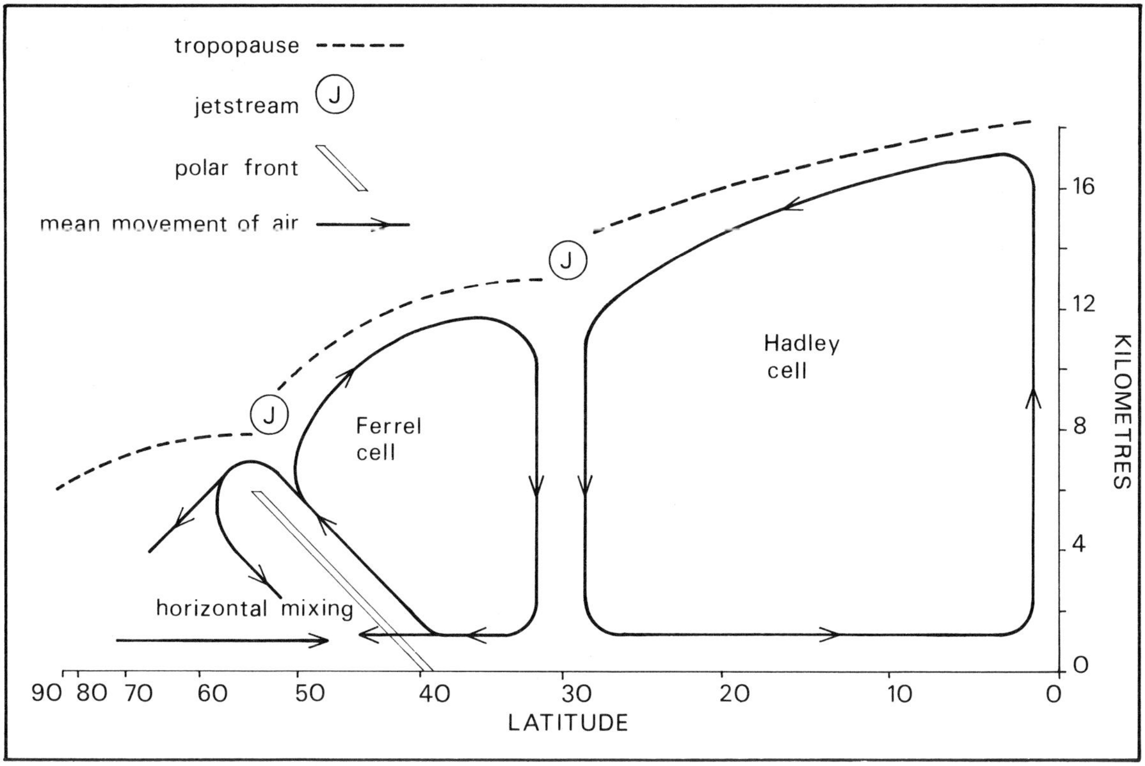

Figure 14 A general model of atmospheric circulation

dry, but as it moves towards the poles, the air becomes increasingly humid. The air of polar origin is, of course, cold, and because of this is unable to hold much moisture. Therefore, at the boundary between the tropical air and the polar air, there is a contest. This boundary is called the Polar Front.

Figure 14 shows the way in which the Ferrel Cell fits into the general pattern of atmospheric circulation. The Ferrel Cell was proposed by W. Ferrel in 1856. The existence of a strong current of air is now questioned by many climatologists, but some features of the Ferrel Cell do exist as indicated above.

---

8 Write down the qualities of both the polar and tropical air at the Polar Front, in terms of temperature, density, humidity and stability.

9 Make a copy of Figure 15 and on it show the stable air and the direction of movement of the tropical air.

10 On your diagram show how temperature varies with altitude on either side of the polar front.

On the poleward sides of the polar fronts the amount of energy available for circulation is very small indeed.

Look back to Figure 8 page 12.

11 Between which latitudes is the deficit of energy greater than 200 Langleys?

12 Should this deficit have an effect on the movement of air in high latitudes?

13 From an atlas find the prevailing wind directions in middle and high latitudes.

14 Using the information provided on Figure 14, draw a diagram to show the pattern of air movement on the earth's surface.

It should now be obvious that the pattern of circulation is not governed solely by the differences in energy receipts at different latitudes. Figure 14 and your interpretation of this on the earth's surface are, unfortunately, a gross simplification. Your reference to atlas maps indicated that winds do not blow over great distances in a meridional fashion. The complicating factors are largely due to the rotation of the earth.

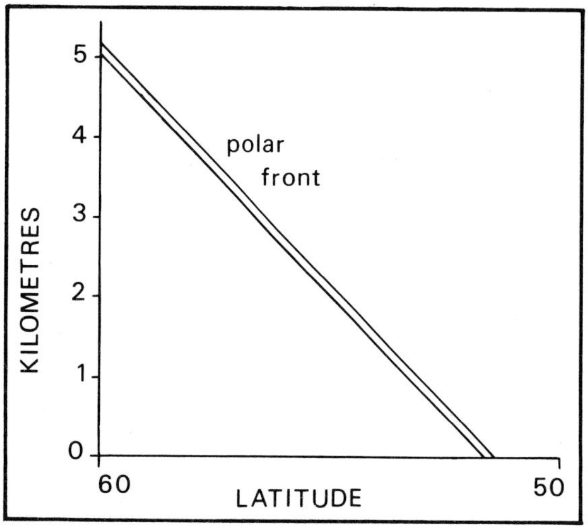

Figure 15 The Polar Front zone

The rotation of the earth

The movement of the earth about its axis has some very important effects on the motion of air on and above the earth's surface. Let us consider the earth as a rotating object.

---

15 (i) The equatorial radius of the earth is approximately 6400km. What is the velocity of a point on the earth's surface at the equator?
(ii) If the surface of the earth at 30°N is 5540 km from the earth's axis, what is the velocity of a point at that latitude?
(iii) If the surface of the earth at 60°N is 3200 km from the axis, what is the velocity of rotation at that latitude?

16 Draw a graph to show the relationship between latitude and the velocity of rotation.

17 Is it possible to sense these variations in velocity?

This apparent difference in velocity at different latitudes may not seem very important — but imagine a parcel of air at the equator at the earth's surface. It is moving very rapidly and if it also starts to move towards either of the poles, it almost instantly moves into latitudes where the velocity of the air is slower.

---

18 Could air leaving the equator at one particular longitude ever reach 30°N at the same longitude? Explain your answer.

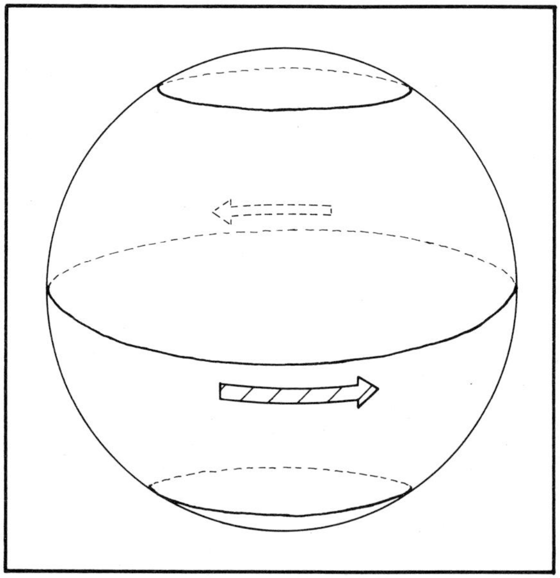

Figure 16 The direction of rotation of the earth

If we take the reverse situation, one of air travelling from high to low latitudes at the earth's surface, it may be seen that a similar pattern emerges. The air at, say 55°N is travelling at 960 km.p.h. while the air at 50°N is travelling at about 1170 km.p.h. A parcel of air travelling from high to low latitudes therefore will not be travelling, or rotating, as fast as the air in the area towards which it is moving.

---

19 Would air setting off from 60°S towards the equator at longitude 40°E ever reach the equator at 40°E? Explain your answer.

So far, so good! What you have shown is that air at the earth's surface moving along a meridian is deflected from its original path by the rotational effect of the earth. One problem remains. In order to work out the direction of apparent deflection you must know the direction of rotation of the earth.

---

20 Make two copies of Figure 16. Label the first "The velocity of the earth's surface at different latitudes," and the second "The directions of apparent deflection of air moving between the poles and the equator."
(i) On the first diagram indicate the various velocities calculated for the particular latitudes in Question 15, bearing in mind

that the values obtained will apply to both northern and southern hemispheres.

(ii) On the second diagram draw in arrows to show the apparent deflection of the air as it moves between the poles and the equator.

21 What do you notice about the direction of deflection of air in both hemispheres whether it is moving towards or away from the equator?

You now have two very important pieces of information in diagram form. Firstly, there is the diagram (Question 14) of the pattern circulation on a non-rotating earth, and secondly that which shows the apparent deflection of air as it moves latitudinally on a rotating earth (Question 20(ii)).

---

22 Draw a diagram to show the pattern of circulation expected on the surface of a rotating earth by combining the above two diagrams.

23 Is this a realistic pattern of circulation? If not, does it represent a good model? Name some factors which may distort this pattern, and explain the ways in which they operate.

This apparent deflection of air is called the **Coriolis Force**, after G. G. de Coriolis who noted the effect in 1835. There are, however, many other forces governing atmospheric motion. The following are some of the more significant ones.

The pressure gradient force

Air will always move from areas of high pressure to areas of low pressure as a result of the **pressure gradient force.** The strength of the pressure gradient force is governed by the rate of change in air pressure between any two points.

Examine Figure 17, which shows two different patterns of pressure — indicated by isobars. Differences in pressure between two points are reduced as air moves away from areas of high pressure (a process called divergence), towards areas of low pressure (a process called convergence).

---

24 Which section, P or Q in Figure 17, has the steepest pressure gradient? Explain your answer.

25 In which section would you expect the winds to be strongest?

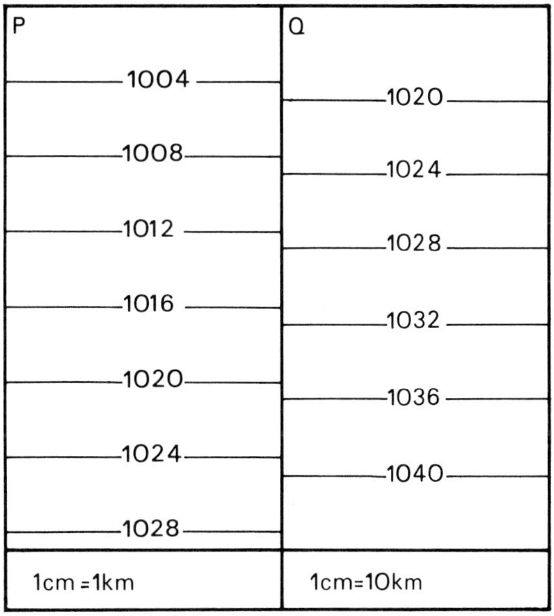

Figure 17 Pressure gradients

It would be reasonable to expect that as air moves from high to low pressure, it would move directly across the isobars with the pressure gradient — in much the same way as an object will roll downhill following the gradient and crossing the contours in a perpendicular fashion. If the earth were not rotating the air would probably move directly across the isobars.

---

26 Make a copy of Figure 14 and on it mark points of convergence and divergence at the earth's surface as high and low pressure areas. What will the pressure situations be in the upper atmosphere directly above surface divergence and convergence?

27 Describe the pattern of high and low pressure you have produced.

28 Compare this pattern with one shown in an atlas and attempt to describe and explain any differences you find (Figure 13 may provide some useful clues).

As has been shown, air does not move in straight lines; it experiences a deflection to the right in the northern hemisphere and to the left in the southern hemisphere. We know that air moves from high to low pressure, but in so doing, it will be immediately deflected from its original course by the Coriolis Force.

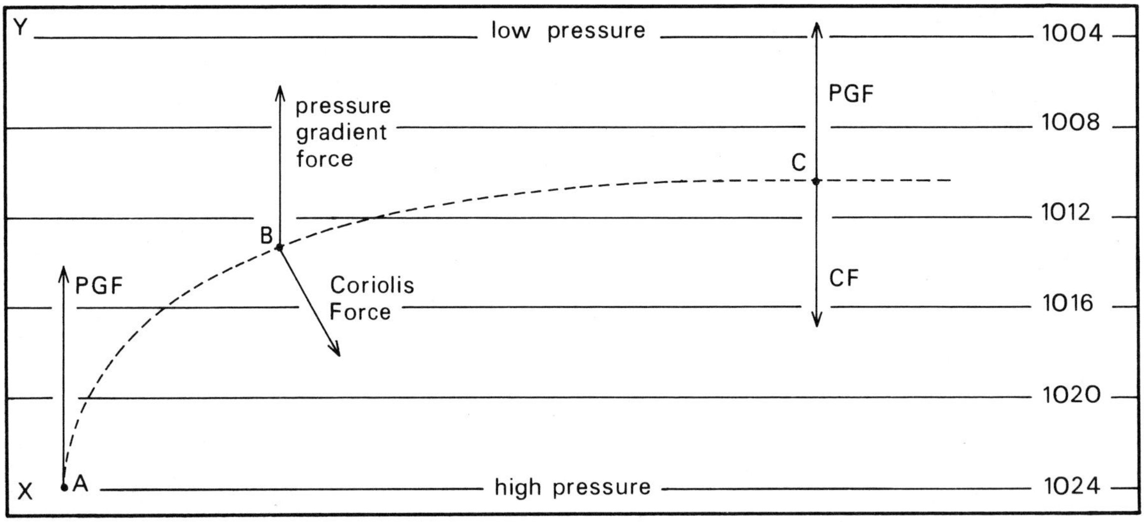

Figure 18 Forces acting on moving air

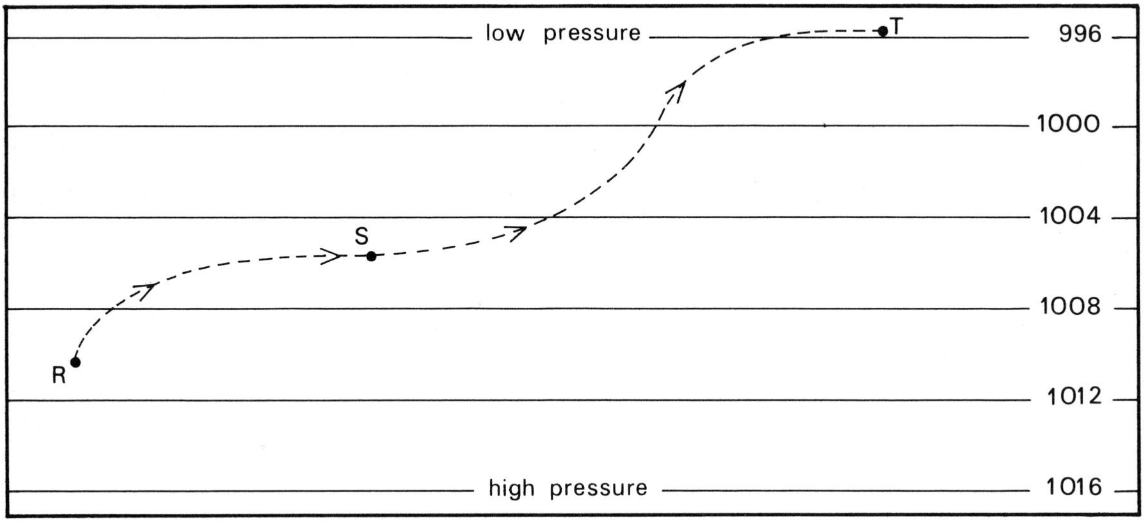

Figure 19

Examine Figure 18 carefully. Air at A moves because of the pressure gradient force — the result of the difference in pressure between X and Y. As soon as it starts to move it is deflected to B because of the Coriolis Force. As long as there is a movement of air from high to low pressure, the Coriolis Force will cause an apparent deflection toward the right in the northern hemisphere, and toward the left in the southern hemisphere. Eventually, at C, a balance is achieved between the deflection due to the Coriolis force, and the movement produced by the pressure gradient force. This balance occurs because as the pressure gradient force increases, so too must the Coriolis Force, and this returns the air to a course parallel to the isobars. In the absence of other effects, air moves parallel to the isobars.

---

29 Make a copy of Figure 19. Annotate the diagram to explain what is happening to the pressure gradient force and the Coriolis Force at locations R, S and T.

This tendency to move parallel to the isobars is only present in the upper atmosphere where there is virtually no friction with the earth's surface. This type of motion is called **Geostrophic flow** and is present where no other forces are operating. At and near the earth's surface, usually below 500 metres, friction comes into play. This reduces the velocity of the air and disrupts the pattern of geostrophic flow. We have already noted that the Coriolis Force increases in strength as the velocity of the air upon which it is acting increases. Friction will reduce the velocity of the air in the geostrophic wind and will therefore reduce the strength of the Coriolis Force.

30 Make a copy of Figure 18 and add to it an arrow indicating the frictional force which slows down the movement of the air.
31 Add to your copy of Figure 18 an arrow which indicates the direction of air movement at or near to the earth's surface where the slowing effect of friction is of importance.

The movement of air near to the surface is seldom parallel to the isobars. Below 500 metres or so, the air usually crosses the isobars at an angle of between 10° and 40° depending upon the velocity of the air and the magnitude of the frictional drag.

Permanent and semi-permanent pressure cells

The circulation of air in the Hadley and Ferrel Cells produces — as you have shown in your copy of Figure 14 — high pressure where the air descends and low pressure where the air ascends. These pressure cells tend to be permanent over the oceans and semi-permanent over the land. The high pressure cells at about 30°N and S are particularly important as they influence weather patterns over much of the earth's surface.

32 With the help of your atlas, explain why the high pressure cells influence the climate of such a large part of the earth's surface.
33 Draw a diagram to show the distribution of pressure on an earth with a uniform surface.
34 Describe and explain the differences between the actual pressure distribution shown in your atlas and the one you have drawn.
35 Briefly outline with the help of your atlas, the apparent relationship between climate and atmospheric pressure.

Thermal winds and jetstreams

These winds are present in the upper atmosphere as a result of temperature and pressure differences at the tropopause. **Thermal winds** are the general movement of air as the result of temperature and pressure differences over large distances. The **jetstream** is a more localised feature being the product of more abrupt temperature and pressure differences. It may be assumed that air at the top of the troposphere is of roughly constant temperature and pressure. The troposphere is not, however, of uniform depth — being some 18km deep at the equator and 6km deep at the poles.
Examine Figure 20 carefully.

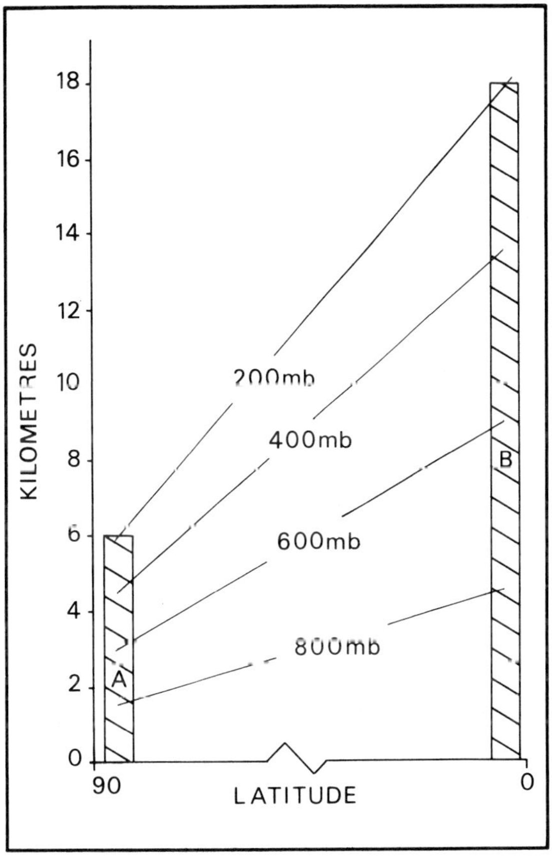

Figure 20 Thermal winds

36 At an altitude of 6km what will the air pressure be in columns A and B?
37 On a copy of Figure 20 indicate the movement of air (a) without, and (b) with the influence of the Coriolis Force.

The thermal wind is generally a westerly current in the upper atmosphere. Its movement tends to be in a direction other than that of the geostrophic flow, and is due largely to the temperature and therefore pressure differences at the same altitude in the troposphere. At certain points in the troposphere, there are discontinuities in the tropopause, with marked thermal contrasts over short distances giving rise to jetstreams. In Figure 14 (p.17) you will note that there are two major discontinuities in the tropopause at about 30° and 55°.

38 Make a large copy of the section of Figure 14 between 20° and 40°, and using information provided on Figure 3, add the approximate temperature and pressure at the tropopause. Explain why a jetstream is found at this latitude.

The jetstreams are concentrated belts of the thermal wind, with average summer velocities of about 100km.p.h. and winter velocities of about 180km.p.h. Occasionally velocities in excess of 350km.p.h. have been recorded. The thermal winds and jetstreams tend to be stronger in winter because the temperature difference between the poles and equator is greater then. The velocity of the jetstream is very high compared with the winds at the earth's surface, but at the average altitude of 12km there is little air and so the force of the jetstream is limited. In terms of their influence upon weather, jetstreams are of great importance because not only do they help to produce depressions and anticyclones in mid-latitudes, but they are also largely responsible for the monsoon circulation in the Indian Ocean basin.

One step further

39 Describe and explain atmospheric circulation.
40 Find out the meanings of the following terms:
(i) vorticity
(ii) Rossby Waves
(iii) cyclonic curvature.
41 Attempt to draw a diagram in three dimensions showing the major features of atmospheric circulation.

Oceanic movements

The forces which govern atmospheric circulation also govern oceanic circulation, except that in the oceans the pressure gradient force is of little significance.

42 With the help of your atlas, outline the apparent relationship between atmospheric and oceanic circulation.

The movement of water in the oceans is similar to the movement of air in the atmosphere in that it is three dimensional.

There are places where water descends to the deep ocean basins and other places where water rises to the surface. Where winds are offshore, the frictional effect of the wind produces a net movement of water away from the shore at the surface. This upwelling of water from the deep ocean basins brings cold water rich in nutrients to the surface (see Figure 21).

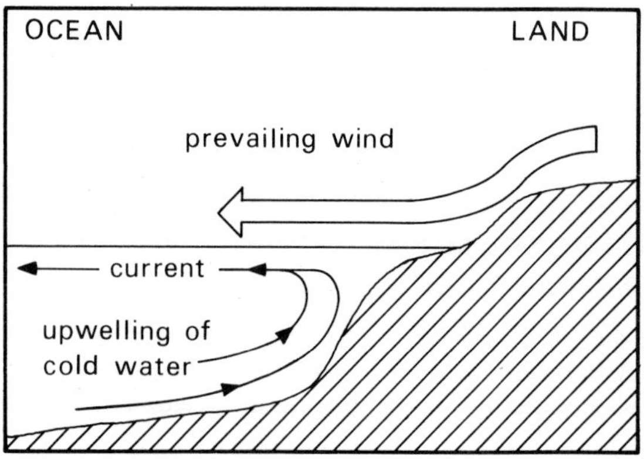

Figure 21 Ocean upwelling

43 With the help of your atlas, suggest two places where such upwelling of water might take place.
44 Why is this movement of water important to humans?

This upwelling of water is called divergence and must be complemented by the sinking or convergence of water in other parts of the ocean basin.
Examine Figure 22.

45 Describe the patterns of ocean currents shown for January and July.
46 Attempt to explain any differences which you have noted.
47 On an outline map of the world mark and name the average positions of the warm and cold

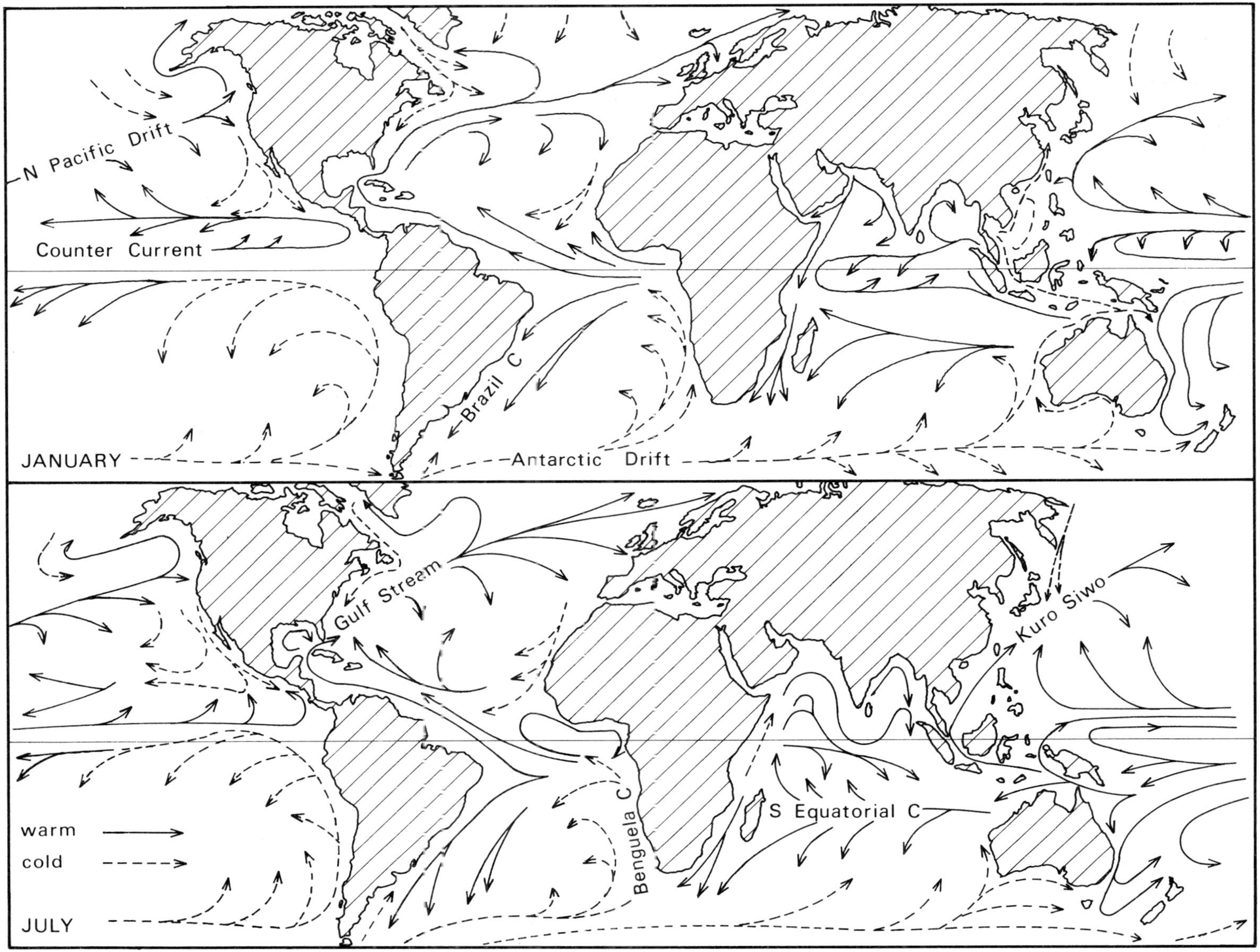

Figure 22 Ocean currents

currents. Using your atlas, add the names of currents not given in Figure 22.

The oceans and energy transfer

Water is more dense than air and as a result is able to hold more heat than air. However, water requires about five times as much heat as air to raise its temperature by the same amount. This has some very important implications. Firstly, and most simply, it explains why on a summer's day in England, the air temperature may be 22°C while the sea is only 12°C.

48 At what time of the year will the seas around Britain be at their warmest?

49 Will the oceans cool down as rapidly as the land in winter?

50 Turn back to Figure 13 (p.16) and explain why the ITCZ does not move as many degrees of latitude over the oceans as it does over the land.

In the section on the Coriolis Force we learnt that as objects move towards the poles from the equator, their rotational velocity, relative to that of the latitude increases. This has important effects upon ocean currents since those which are moving away from the equator are well defined, i.e. easily measured, because of their velocity, e.g. Kuro Siwo and Florida Current, both of which achieve velocities of about 5·5km.p.h. The ocean current influencing the coasts of North Western Europe, the North Atlantic Drift, travels at a velocity of between 5 and 8 kilometres per day.

51 Do you think that cold currents returning to the equator will be particularly well defined? Explain your answer.

Insolation and the mixing of the surface layers, produce a pronounced heating effect to a depth of 10 metres in the oceans. This compares with a depth of only half a metre for the heating of the land. However, in warm ocean currents the body of warm water may be as deep as 200 metres, although a depth of 50 to 100 metres would be

Figure 23 Temperature anomalies from average for latitude

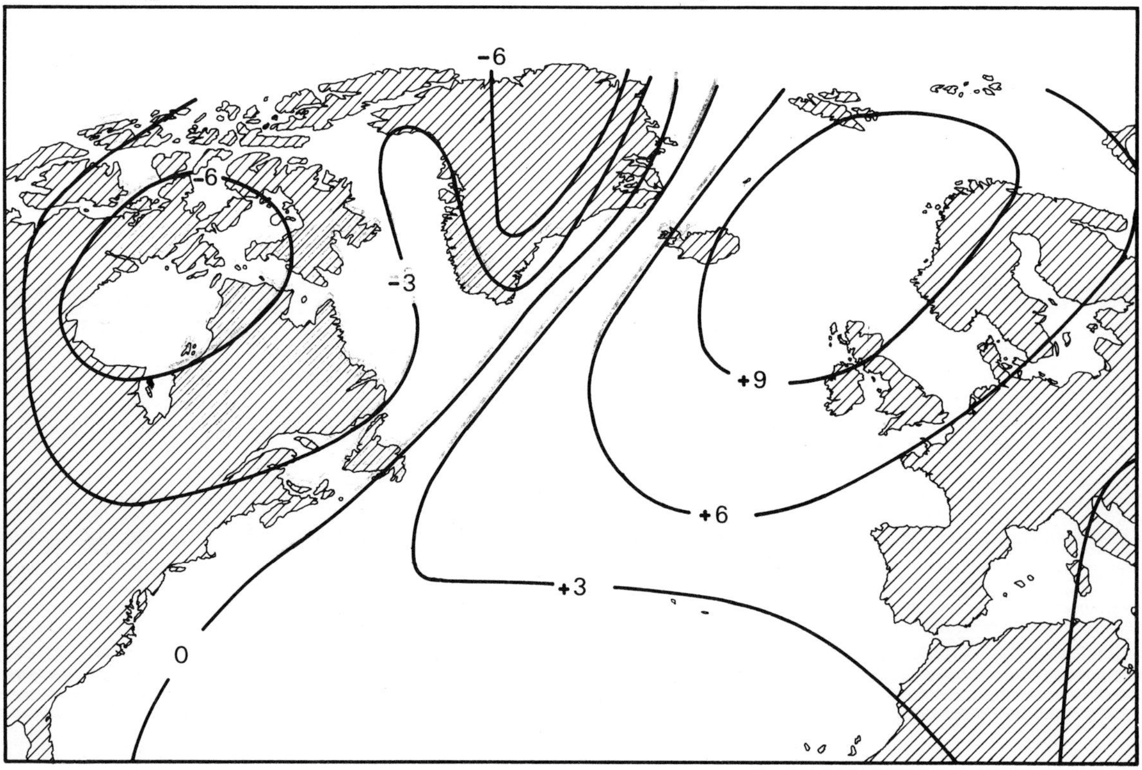

more common. The temperature of the ocean current and its relatively slow cooling and heating have marked effects on many climates.
Examine Figure 23.

---

52 With the help of Figure 22, account for the pattern of air temperature shown in Figure 23.

53 Figure 23 shows temperature anomalies from the normal for the latitude. With reference to your atlas, name one cold and one warm ocean current which could produce a similar effect to that shown in Figure 23.

Chapters 1, 2 and 3 have outlined the source of energy and the means by which it is transferred to other parts of the earth's surface. In other words, they have examined the mechanics of climate production. The next three chapters examine the outcome of the major influences upon climates in low, middle and high latitudes.

4 Climates of Low Latitudes

Low latitudes, those lying between 30°N and 30°S of the equator, contain high energy climates. The sun is never lower than an angle of about 35° at midday, and is directly overhead at many places once or twice a year. This is in sharp contrast with Britain where minimum elevation of the midday sun is about 15° and the maximum elevation is little more than 60°.

Look back to Figure 8 (p.12).

---

1 Why does the latitude scale not follow a regular pattern?
2 Assess the importance of low latitudes in terms of global energy receipts.
3 Write a statement about the climates of low latitudes in terms of energy.

We have already noted the major features of pressure and circulation in low latitudes. We have noted, too, that these features are largely permanent, e.g. the Hadley Cells, the subtropical high pressure cells and the Inter Tropical Convergence Zone (ITCZ). Yet we see on television and read in newspapers about weather "disasters" in low latitudes — the drought in sub-Saharan Africa of the 1980s, and the devastation of Darwin, Australia by Cyclone Tracey in 1974 are two examples. If the features which produce the climates of low latitudes are permanent, why do these newsworthy abnormalities occur? In reality, most tropical climates are highly variable and although disasters occur they are not commonplace. We shall now look at three low latitude climates.

The equatorial climate

This climate contains the features which most people assume to be characteristic of the tropics. It is a hot and wet climate, produced largely by the ITCZ.

Examine Figure 24.

Figure 24 The Hadley Cell

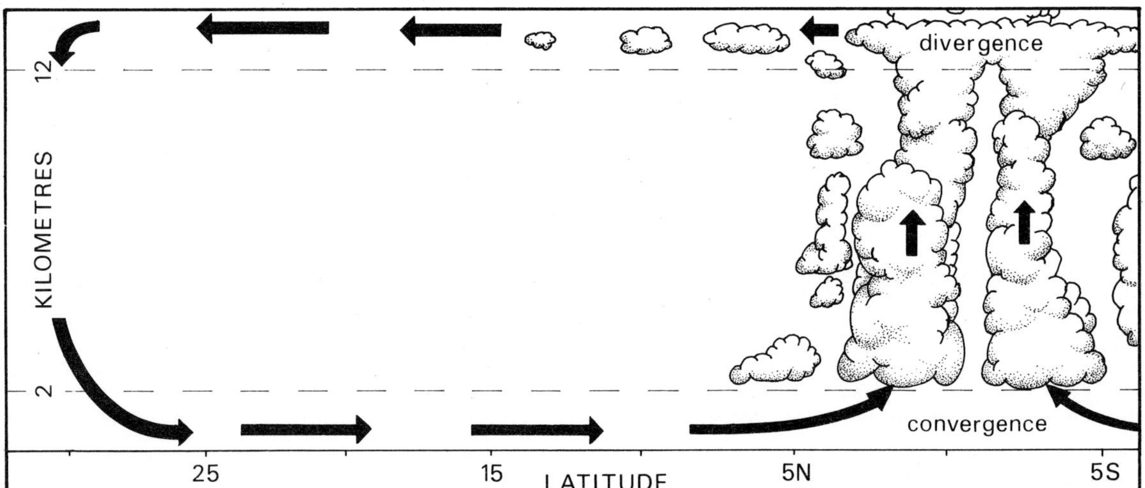

4 Make a copy of Figure 24 and annotate it, explaining what happens as the air ascends.
5 As the ITCZ is not static, what result will this have in climatic terms in low latitudes? (Figure 13 (p.16) may also provide additional information.)

The ITCZ is produced where the North-East and South-East Trade Winds converge. There is not a front at the meeting of these two types of air as they both have similar characteristics. At the ITCZ the air rises rapidly, producing clouds up to an altitude of 12km with great turbulence. The photograph shows cloud associated with the ITCZ.

Examine the climate details given for Balikpapan, Indonesia (1° 17′S, 116° 51′E) in Table 2.

Table 2: Climate details for Balikpapan, Indonesia, 1° 17′S 116° 51′E

	J	F	M	A	M	J	J	A	S	O	N	D
°C	26	26·5	26·5	26	26	26	25·5	26	26	26	26	26
mm	201	175	231	208	231	193	180	163	140	132	168	206

6 Produce a climate graph using the data given.
7 On the graph, write the annual temperature range and the total precipitation.

Figure 25 Hythergraph for In Salah

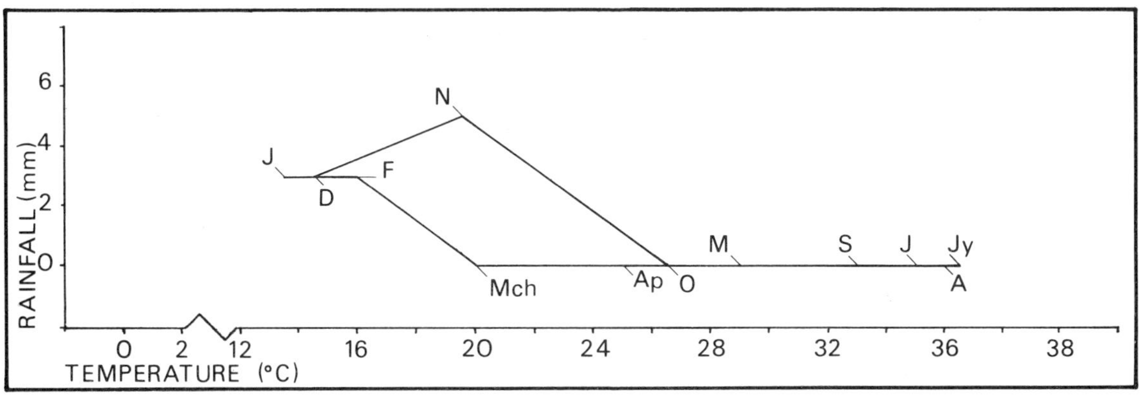

Balikpapan has not recorded a temperature of less than 19°C nor one higher than 33°C. The driest months are September and October with rain falling on 9 days. The wettest months are March and December with rain falling on 15 days.

8 From the details provided, briefly describe the nature of the equatorial climate.
9 Would you expect some months to be hotter and/or wetter than others? Explain your answer.
10 Using an atlas, attempt to explain the distribution of the equatorial climate.

The hot deserts

Looking at Figure 14 (p.17) you can see the ascent of air in the ITCZ. This rising air has lost much of its water vapour.

11 What will the humidity of the air be at the tropopause in the Hadley Cell?
12 Why is air able to expand as it ascends? What happens to it as it descends?
13 What will be the nature of the air in the Hadley Cell as it reaches the surface of the earth at about 30°N and S°?

The air descending at about 30°N and S° forms the sub-tropical high pressure cells. Air moves outwards — diverges — from these high pressure cells. The air moving towards the equator forms the Trade Winds, and that moving towards the poles forms the Midlatitude Westerlies.

14 With the help of an atlas, describe the relationship between the hot deserts and the Trade Winds.
Look at Figure 25. It contains the climate details for In Salah, Algeria, 27° 12′N, 2° 28′E.

Photo showing ITCZ and clear skies over Africa

15 Using the same axes, with appropriate scales, plot the data for Balikpapan (Table 2) in a similar form.
16 What are the advantages and disadvantages of this method of plotting climate data?

The climate of In Salah is dominated by the descending dry air of the subtropical high pressure cells. The photograph shows the clear skies over the Sahara, where air is descending from the upper atmosphere. Its average annual rainfall is 17mm, and some months are totally dry — for 15 consecutive years no rainfall has been recorded in the month of July. The maximum shade temperature recorded is 50°C, in the months of June, July and August. The lowest temperature ever recorded is −3°C. These are extremes, but in July the average maximum temperature is 45°C and the average minimum is 6°C in January.

These figures are by no means unusual for deserts. Death Valley, California 36° 28′N, 116° 51′W, has a maximum temperature of 57°C and a minimum of −9°C. Alice Springs, Australia 23° 38′S, 135° 58′E has a maximum temperature of 44°C and a minimum of −7°C. These are extremes again, but it must be emphasised that the hot deserts are very hot indeed. Marble Bar, W. Australia 21° 9′S, 119° 4′E is one of the hottest places on earth in that temperatures in excess of 38°C have been recorded for 160 consecutive days.

If it is so hot in the daytime, why is it so cold at night? The answer is quite simple and may be found in Figures 4 and 7 (p.7 and p.11).

---

17 Assess the role and importance of clouds in helping to maintain high temperatures at night.
18 Draw two diagrams, and annotate them, one to explain why deserts in low latitudes are hot during the day, and the other to show why they are cold at night.

Not all hot deserts are hot! How could this be when they generally lie within about 30° of the equator? Figures 21 and 22 (p.22 and p.23) reveal why.

19 Along which sides of the continents are there cold ocean currents between 35°N and S of the equator?
20 Are the winds onshore or offshore between 35°N and S?
21 Examine Figure 26 and explain why there are lower than average temperatures in the areas shown.

Walvis Bay, Namibia, 22° 56′S, 14° 30′E, has a modified desert climate. Despite its latitude, it has never experienced a temperature of more than 40°C, and the average maximum temperature varies between 19°C in September and October and 24°C in April. The coldest months are July and August with average minimum temperatures of 8°C. The lowest temperature ever recorded is 4°C. Despite this, it is most definitely a desert with only 22mm of rain in a year. Compare that total with the monthly totals for Balikpapan (Table 2).

Figure 26 Temperature anomalies in low latitudes

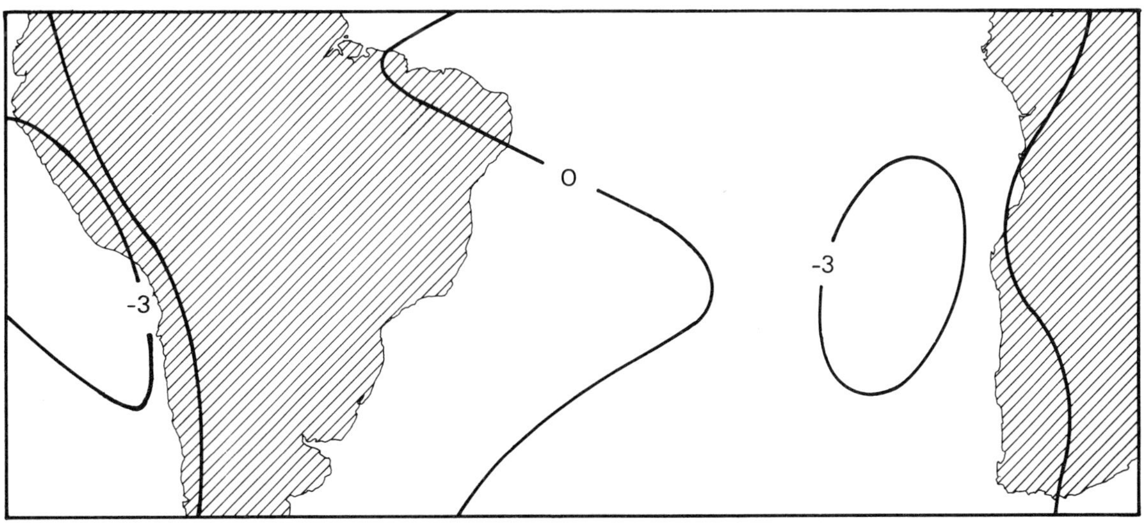

22 Use an atlas to tell you how much precipitation is received by areas which are called deserts. Can deserts be defined solely in terms of precipitation totals?
What other factors could be involved?

Tropical wet and dry climates

Within the tropics there are two climates which separately display extremes of humidity and aridity and which generally have high temperatures. These two climates are, as we have seen, produced by the ITCZ and the subtropical high pressure cells with their associated Trade Winds. But we have also seen that the ITCZ is not stationary, and that the intensity of the subtropical high pressure cells varies. The ITCZ "migrates" with the overhead position of the sun, taking with it the very heavy convectional rainfall. Similarly, the subtropical high pressure cells extend themselves towards the equator or towards the poles at certain times of the year.

23 What will the effect of this movement be upon the areas which lie between the equatorial and desert climate zones?

24 Compare Figure 27 with Figure 13 (p.16). Describe and explain the relationship between the information given in both diagrams.

25 With the help of an atlas, see if the same relationship holds true for South America.

26 Describe the pattern of precipitation in the areas of seasonal precipitation shown in Figure 27. Will precipitation totals be uniform throughout these areas?
Examine the data given in Table 3.

27 Describe the differences between the two sets of data, and state which set of figures applies to a location in the north of Africa and which to one in the south.

Tropical wet and dry climates are divided into two different types — monsoon and savanna. As you have probably noted, the ITCZ is unable to bring an even amount of precipitation to the areas over which it passes, and therefore some areas receive comparatively small amounts of rain,

Figure 27 Seasonality and totals of precipitation

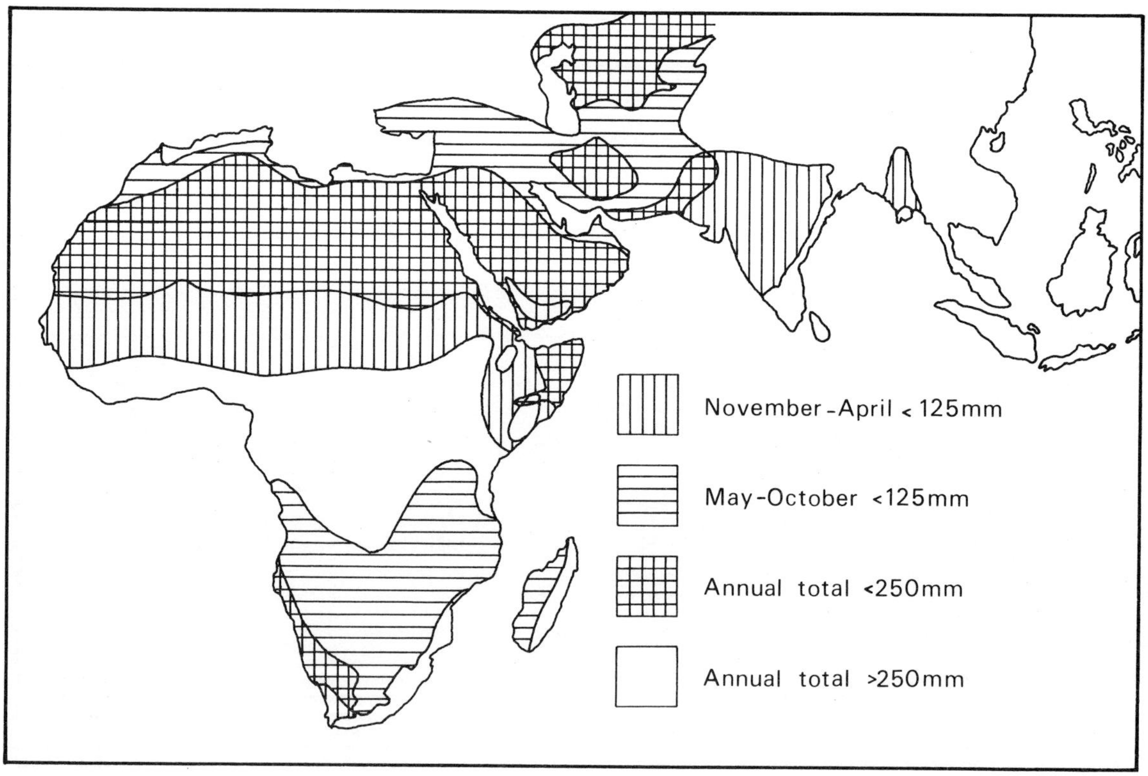

Location		J	F	M	A	M	J	J	A	S	O	N	D
A	°C	23	24	25·5	25·5	25·5	27·5	26·5	26	27	27	25	23·5
	mm	3	3	0	0	10	58	282	500	310	109	18	3
B	°C	22·5	22·5	21	20	17·5	14·5	15	17	21	23·5	23	22
	mm	351	264	234	33	3	0	0	0	0	18	140	252

Table 3: Tropical wet and dry climates in Africa

whilst others receive large amounts in their wet seasons. The total precipitation defines the climate, but of greater importance is the proportion of the precipitation falling in the wet season.

---

28 Describe and explain equatorial, savanna, and monsoon climates as shown in Figure 28.

29 Using Figure 28, classify the data given in Table 4 into monsoon or savanna varieties of the tropical wet and dry climate.

The seasons of tropical climates may often be given the same names as ours, but they tend to be defined in a different way. We know that in mid-latitudes the summers are markedly warmer than the winters. In low latitudes this is not necessarily the case.

---

30 How are seasons defined in low latitudes?

The climates of low latitudes have one common characteristic — their high energy. The desert and equatorial climates display little variability, being almost without exception, dry or wet respectively. The monsoon and savanna types have characteristics of both the desert and equatorial climates, because they are transitional between the two. They have wet and dry seasons. Average temperatures in low latitudes are seldom below 20°C, except in mountainous regions or where cold ocean currents modify the climates along the coasts as occurs in Chile and Namibia.

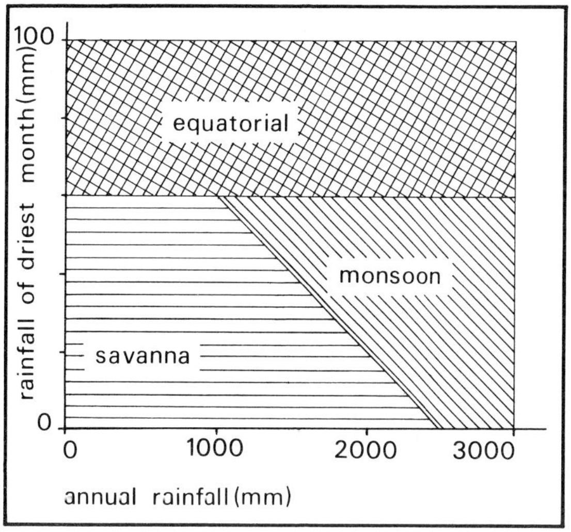

Figure 28 Definition of humid tropical climates

It would appear that the climates of low latitudes are largely predictable. Unfortunately for the inhabitants of these areas the climates do display large scale local variations — some of which are disastrous for humans, livestock and crops. One of these variations is examined in some detail in the following case study.

Table 4: Precipitation in tropical wet and dry climates

Location	J	F	M	A	M	J	J	A	S	O	N	D
Jakarta 6° 11′S 106° 50′E	300	300	211	147	114	97	64	43	66	112	142	203
Conakry 9° 31′N 13° 43′W	3	3	10	23	158	559	1298	1054	683	371	122	10
Darwin 12° 28′S 130° 51′E	386	312	254	97	15	3	0	3	13	51	119	239
Antananarivo 18° 55′S 47° 33′E	300	279	178	53	18	8	8	10	18	61	135	287

Tropical cyclones — a case study

The amount of energy available in low latitudes is admirably demonstrated by the tropical cyclone. A tropical cyclone is certainly an aberration, since only about fifty occur each year. The intensity of these cyclones varies, some may cause great damage, but all bring heavy rain and winds in excess of 120km.p.h. There are different local names for these features of atmospheric circulation; in the Atlantic they are called **hurricanes**; in the Indian Ocean, **cyclones**; and in the South China Sea, **typhoons**.

The formation of tropical cyclones

Tropical cyclones originate over the warm tropical oceans where the surface temperature of the water is in excess of 27°C (see Figure 29). This very warm water will evaporate and saturate the air, but this alone is insufficient to produce a cyclone.

---

31 What normally happens to the moist air over the tropical oceans?

32 With reference to Figure 24 (p.26), state what conditions are necessary to draw this moist air into the upper atmosphere.

33 Describe conditions of air pressure in the upper and lower atmosphere.

One of the main conditions necessary for the development of tropical cyclones is the **Easterly Wave**. This is an area of low pressure which is basically an extension of the ITCZ lying roughly north to south in the Trade Wind belt. These troughs of low pressure are most common in the summer season. The low pressure of the Easterly Wave often breaks down the upper air flow, allowing the moist air to rise into the upper atmosphere. Only a few Easterly Waves per year ever have sufficient intensity to develop into tropical cyclones.

Examine Figure 30.

---

34 Describe the form of a tropical cyclone.

The spiralling pattern of winds in the tropical cyclone, as shown in the photograph opposite, would not exist if the earth did not rotate. Tropical cyclones very rarely form within five degrees latitude of the equator.

---

35 Why should cyclones not form very near to the equator?

36 Using the information provided so far, state how you would monitor conditions over an area of ocean in order to provide a tropical cyclone warning service.

Figure 29 The origins and paths of tropical cyclones

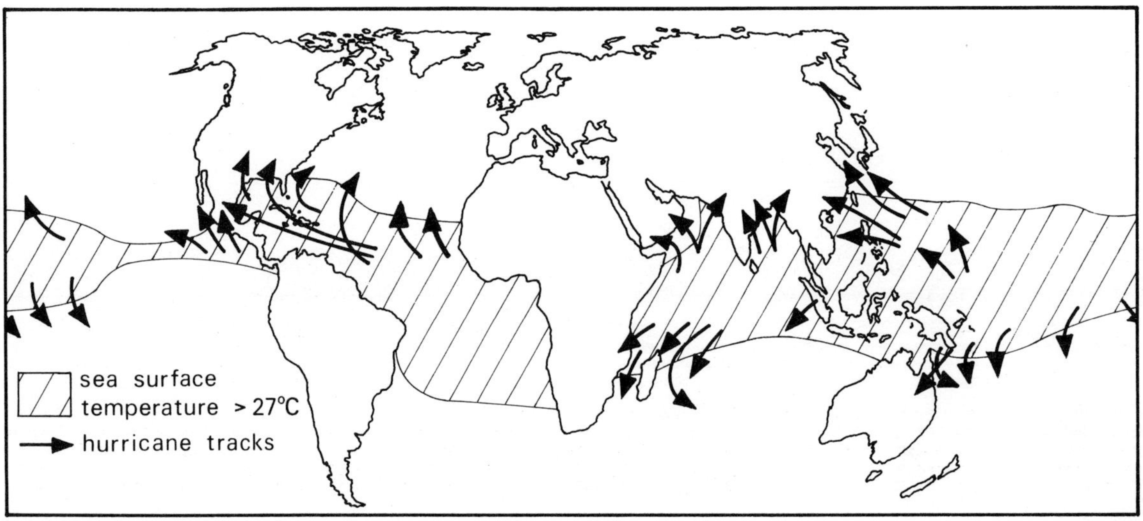

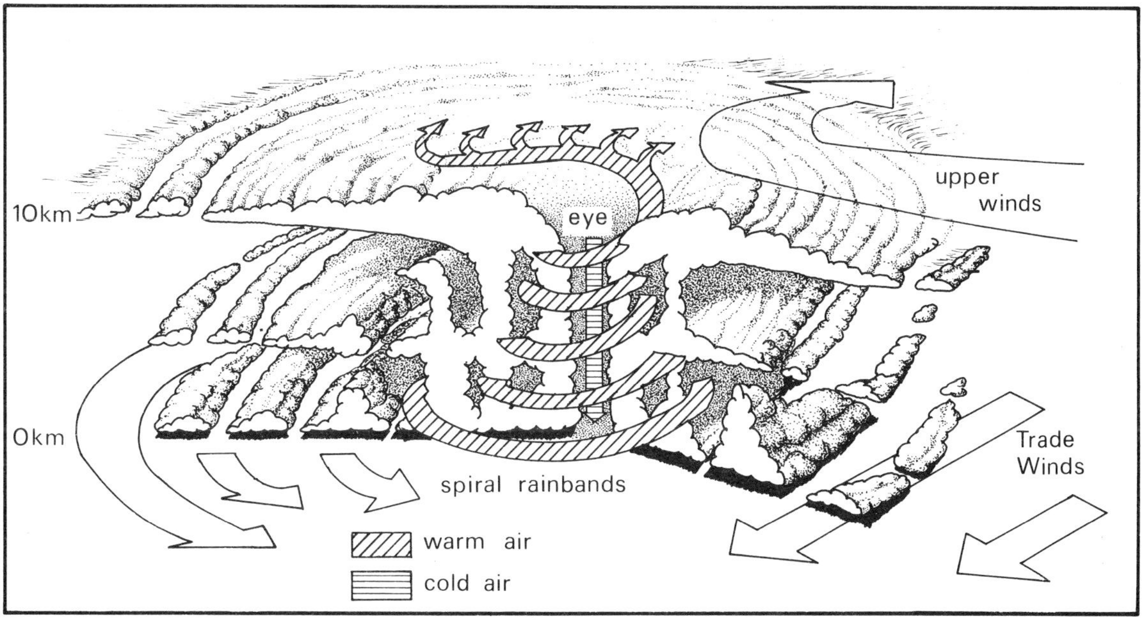

Figure 30 The form of tropical cyclones

Hurricane Debbie

Obviously the intensity of these disturbances varies, and in the U.S.A. and Canada they are classified according to wind speed.

WIND SPEED	NAME OF DISTURBANCE
<17 m/s	tropical depression
18–33 m/s	tropical storm
>34 m/s	hurricane

Areas affected by tropical cyclones
Sometimes tropical cyclones are particularly intense and are able to travel great distances outside the tropics. As tropical cyclones travel along in the Trade Wind belt, their paths are usually deflected towards the poles first, and then in a more easterly direction (see Figure 29). The reason for these changes in direction lie in the movement of air in the upper atmosphere.

---

37 Bearing in mind the conditions necessary for the formation of tropical cyclones, why is it remarkable that they are able to travel outside of the tropics?

38 Describe and explain the distribution of areas influenced by tropical cyclones.

39 Attempt an explanation of the relationship between the upper atmospheric flow and the tracks of tropical cyclones.

As tropical cyclones travel away from the equator or further away from the sea, they generally decrease in intensity. However, there are cases of tropical cyclones travelling great distances, as in 1954 when on October 15th a hurricane, which had travelled from a position north of Cuba on October 14th, caused more than 80 deaths and 24 million dollars worth of damage in the Toronto area of the north shore of Lake Ontario. Winds of 115km.p.h. were recorded, and 178mm of rain fell to the northwest of Toronto. There have been cases of tropical cyclones travelling as far north as Halifax, Nova Scotia 44° 38′N. Most tropical cyclones are restricted to tropical or sub-tropical latitudes.

The life cycle of tropical cyclones
The major source of energy of a tropical cyclone is the moisture evaporated from the tropical oceans. With the presence of a surface low or convergence, the warm moist air is drawn into the upper atmosphere. This moist air cools giving first a condensation of water vapour to produce clouds and then precipitation. Under normal circumstances, these convection cells never have sufficient energy to disrupt the upper air flow. The presence of an Easterly Wave alters the situation. The intensity of the low pressure area of the Easterly Wave enhances the upward movement of the moist air and it is often able to break into the upper air flow. The tropical storm travels along in the Trade Wind belt until it has broken into the upper westerly air flow, which is removing air from the top of the cyclone so that more may be drawn in at the base. Once it has broken into the upper westerlies it will begin an eastward movement, encircling the centres of the sub-tropical high pressure cells.

Once the tropical cyclone has moved away from the areas where the oceans have surface temperatures in excess of 27°C, their source of energy, the water vapour from the tropical seas, is removed. However, as the cyclone moves away from the equator, the strength of the Coriolis effect increases. The velocity of the storm therefore increases, while at the same time the amount of energy available to the storm is slowly dwindling.

As the cyclone moves over land, it is immediately cut off from its source of energy and so it will start to decay. Similarly, when the cyclone moves over oceans whose surfaces are well below 27°C, it will receive progressively less water vapour, and so will decay slowly. In either case the tropical cyclone may help to form the basis of a midlatitude cyclone or depression as it brings warmer than usual air to cooler latitudes.

5 Midlatitude Climates

The climates of mid latitudes differ from those of low latitudes, largely as a result of the great fluctuations in the receipts of solar energy over the year. Turn back to Figure 7 (p.11) and examine it carefully.

---

1 Calculate the length of daylight at 30° and 60° latitude on June 21st and December 22nd.
2 Write a brief comparison of the sets of figures you have abstracted from Figure 7.

These marked contrasts mean that when the sun is high or the daylight hours many, temperatures will be correspondingly high. During winter, the low angle of the sun and the short period of daylight ensure lower temperatures. Consequently, in midlatitudes, seasons are defined in terms of temperature, rather than moisture as is the case in the tropics.

In low latitudes, the hot deserts have temperatures in excess of 18°C in most months. The savanna regions have temperature ranges in the order of 12°C, while the equatorial climates have temperature ranges of less than 6°C. In midlatitudes, and in the continental interiors in particular, there may be temperature ranges in excess of 20°C and as much as 40°C.

This chapter now examines three climates from midlatitudes: warm and cool coastal varieties and the climate of the midlatitude continental interiors. For the purposes of this chapter we shall consider midlatitude climates to be those which lie between 30° and 60°N and S of the equator.

The warm temperate western margin or Mediterranean climate

The Mediterranean lands are well known as places of sunshine and exotic crops. Much of the world's wine, citrus fruit and other crops are produced and many holidays taken in these areas. The name Mediterranean is taken from the largest area in the world to display this type of climate. There are smaller areas of the same type of climate in latitudes similar to that of the Mediterranean Sea. Examine the figures given in Table 5.

---

3 Outline the general pattern of the Mediterranean climate, using the data given for the locations A and B.
4 In which hemispheres would you expect to find A and B? Explain your answer.

With such differences between summer and winter months, you would be right to assume that there are two contrasting influences on the climate.

Refer to Figure 27 (p.30).

Table 5: Mediterranean climates

Location		J	F	M	A	M	J	J	A	S	O	N	D
A	°C	23	23	22	19	17	14	13	13	15	17	19	22
	mm	8	10	20	43	130	180	170	145	86	56	20	13
B	°C	8	9	11	14	17	21	24	24	21	17	13	10
	mm	116	85	73	62	44	31	19	32	64	107	147	135

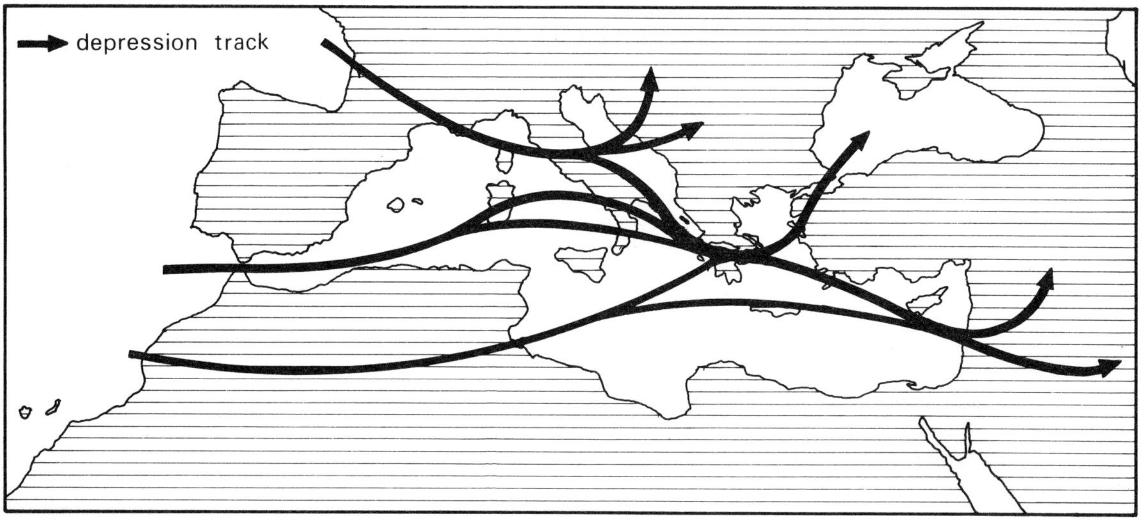

Figure 31 Depression tracks in the Mediterranean

5 Where do the Mediterranean lands lie relative to the desert lands of northern Africa?
6 How does this help to explain the hot dry summers of the Mediterranean climate?

Immediately to the north of the Mediterranean lands the climate is similar to that experienced in Britain — there is no dry season.

7 Apply what you have already learned of the effects of the migrating position of the overhead sun, to an explanation of the Mediterranean type of climate.
8 What problems would such a climate pose for agriculture, and how could they best be overcome?

The Mediterranean climate is transitional between what could be classed as a tropical climate (hot desert) and a temperate climate (cool temperate western margin).

The dryness of the summer months is due to the poleward movement of the sub-tropical high pressure cells, while the wet winters are the result of depressions following tracks some 10° to 15° further south than they do in the winter (see Figure 31).

9 In which ways would natural vegetation have to adapt to the conditions of the Mediterranean climate in order to survive?

Cool temperate western margins

This is the climate for which Britain is notorious. It is not particularly wet, nor is it particularly cold. Its notoriety derives from its unpredictability. Some years are very dry, for example, 1976, 1983 and 1984; the summers may be warm, and may even be

Table 6: Cool temperate western margins

Location		J	F	M	A	M	J	J	A	S	O	N	D
Prince Rupert, Canada 54°18′N 130° 26′W	°C mm	2 214	3 209	4 180	6 184	10 123	12 107	14 121	14 147	12 242	9 359	5 269	3 259
Punta Arenas, Chile 53° 10′S 70° 54′W	°C mm	10 38	10 23	8 33	7 36	4 33	3 41	2 28	4 31	5 23	7 28	8 18	10 36
Oban, Scotland 56° 25′N 5° 30′W	°C mm	4 146	4 109	6 83	8 90	10 72	13 87	14 120	14 116	12 141	10 169	7 146	5 172

less wet than the winters. But this is not always the case, 1980 and 1985 experienced very wet summers. Winters are wet and they are generally mild, despite years like 1947, 1962 and 1985.

These climates are found on the western margins of continents between about 45° and 60° N and S. The high rainfall is the result of many factors. Examine the data given in Table 6.

---

10 Describe the patterns of the climate for the three locations in terms of:
(a) seasonality of precipitation
(b) total precipitation
(c) average temperature
(d) temperature range.

11 Use an atlas to find the locations named in Table 6. Give reasons for the amounts of precipitation received.

Examine the data provided in Table 7.

Table 7: Decreasing precipitation across Europe

LOCATION	ANNUAL PRECIPITATION (mm)
Valentia	1400
Cardiff	1064
Flushing	689
Kassel	631
Leipzig	595

---

12 With the help of an atlas, describe and explain the data given in Table 7.

Proximity to oceans, and humid onshore winds may help to explain the high rainfall of the cool temperate western margins, but much of the precipitation comes from a succession of midlatitude cyclones or depressions. These deserve detailed examination, for like tropical cyclones, they are a major point of energy exchange between the upper and lower atmosphere.

Midlatitude depressions

Midlatitude depressions have a definite life cycle, during which they pass through a number of clearly defined stages. These are **frontogenesis, maturity, occlusion** and **frontolysis**.

Frontogenesis

This is the initiation of the depression where two types of air meet — tropical from lower latitudes and polar from higher latitudes. These air types have different characteristics, and so their meeting point is well defined and is referred to as a front. Over the North Atlantic this is called the Polar Front. The nature of the air on either side of the Polar Front depends upon its area of origin and what type of surface it has passed over, i.e. land/water, ice/hot desert. The positions of the fronts varies throughout the year. Generally they are further south in the northern hemisphere in winter than they are in summer.

Over the Atlantic Ocean, the air on either side of the Polar Front is moving in roughly opposing directions. To the north of the front there is cold easterly air and to the south there is warm westerly air. The front is a particularly unstable area and the balance between the two types of air is easily upset. When this occurs, a wrinkle or corrugation develops in the front, and this moves from west to east.

These wrinkles in the Polar Front were not fully understood until the existence of the jetstream was established after the Second World War. It is the jetstream which provides the means by which the wrinkle in the Polar Front develops into a mature depression. The jetstream at about 50° to 60° N will occasionally be directly over the Polar Front. More important than this, however, is the fact that waves develop in the jetstream, which means that the Polar Front will seldom be parallel to the jetstream. Returning to the Coriolis force (p.19), you will remember that as air moves towards the equator its velocity decreases, and as it moves towards the poles its velocity increases. This effect is largely due to what is called the conservation of angular momentum — an effect often employed by ice skaters as they pull in or spread out their arms in order to increase or decrease their rate of revolution. So as air moves towards the poles in the jetstream its velocity will increase and the opposite will happen as it moves towards the equator. Examine Figure 32.

---

13 Will the velocity of the air entering box P be greater or less than that of air leaving box P? Explain your answer.

14 Similarly with box Q, what will be the velocity of the air leaving compared with that arriving?

Remember that high pressure is associated with

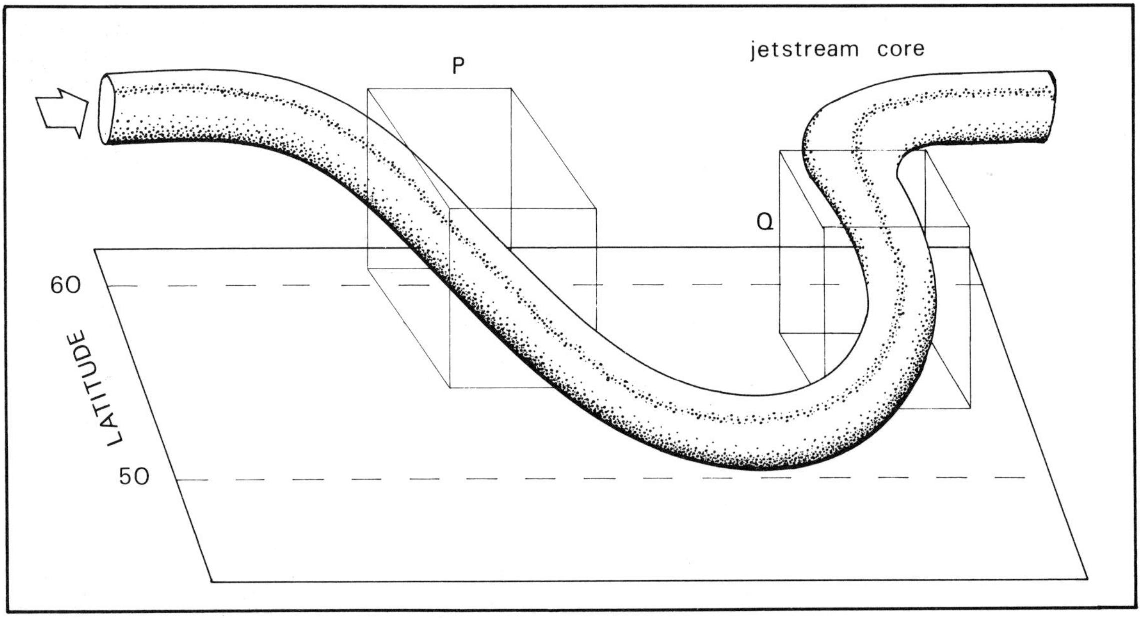

Figure 32 Convergence and divergence in the upper atmosphere

the dispersal of air and low pressure is associated with a drawing in of air (ref. Chapter 3 Question 26 p.19).

15 Draw a vertical cross-section through boxes P and Q. On it indicate which area will experience high and which low pressure in the upper air.

16 Add to your diagram for Question 14 the relationship of this to the air flow at the surface.

Completion of Questions 15 and 16, show that air is being drawn up into box Q. It is precisely this which causes a depression to deepen, developing from a mere wrinkle to a sizeable incursion of warm air into the territory normally occupied by the cold polar air. The low pressure caused by the jetstream will draw warm air up into the atmosphere producing the typical cloud pattern associated with a depression at the warm front (see Figure 33A).

The mature depression

All the features of the next stage of the depression are shown in Figure 33A and B.

Examine Figure 34.

17 Describe and explain the movement of air and the likely transfer of energy as shown in Figure 34.

As the depression progresses through the mature stage, the cold front advances more rapidly than the warm front, undercutting the air of the warm sector. The warm air is gradually forced out of contact with the ground. The cold air has a tendency to remain in contact with the ground as it is more dense than the warm air. The photograph on page 40 shows a mature depression approaching Western Europe.

Examine Figure 35 which shows the sequence of possible changes resulting from the passage of a depression over an area.

18 Describe and explain the changes which would take place if a depression were to pass over the area in which you live (use Figure 33 also).

Occlusion

This stage occurs as the cold front overtakes the warm front. The advancing cold air behind the cold front is more stable than the air of the warm sector, and is therefore able to lift the warm air out of contact with the earth's surface. The warm air cools rapidly eventually producing heavy rain.

Examine Figure 33C, D, and E.

19 Suggest reasons for the difference in air temperature on either side of the warm sector.

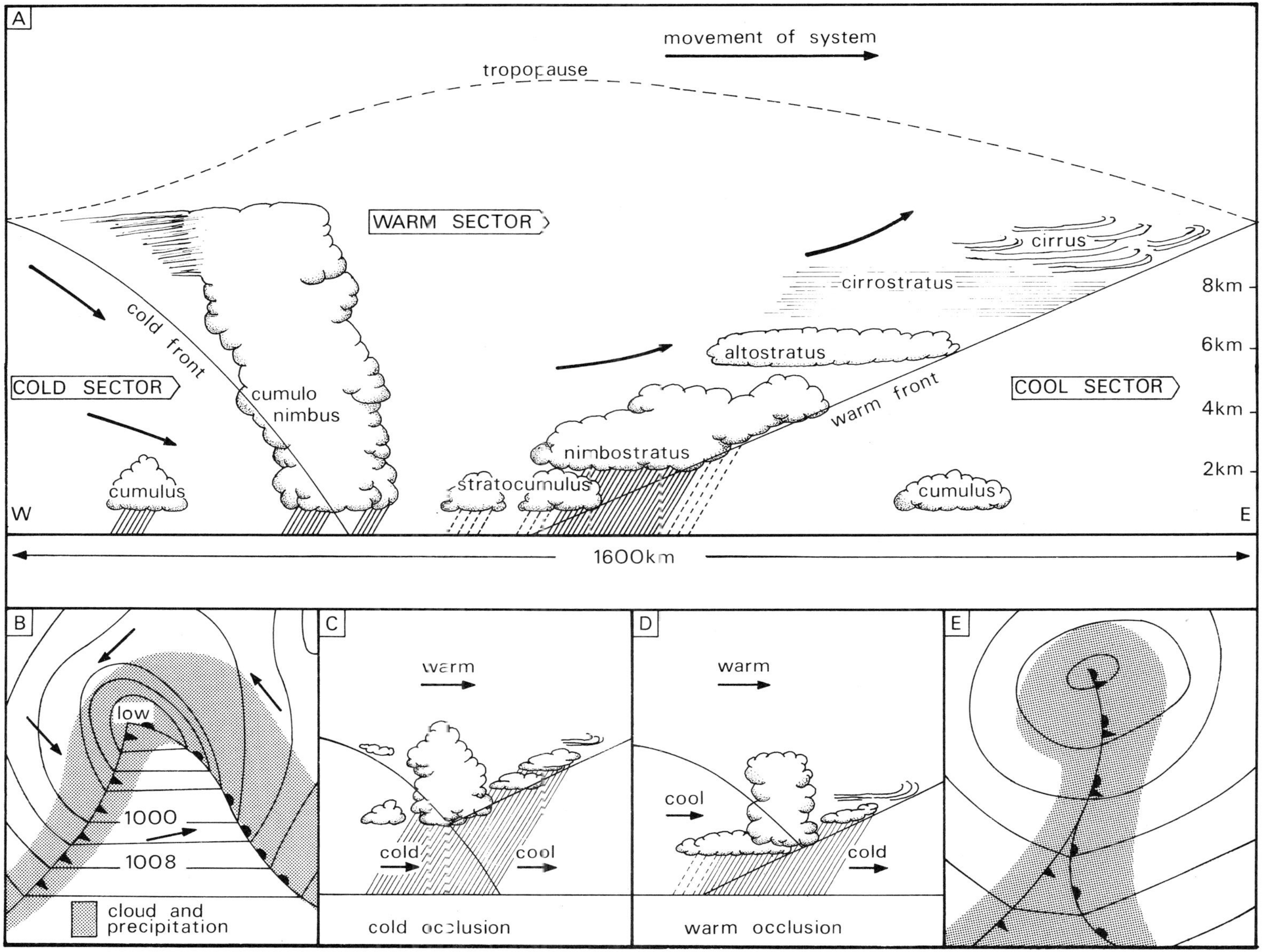

Figure 33 Characteristics of depressions and occlusions

20 Describe and explain the difference between warm and cold occlusions.
21 Make copies of Figures 33B and 33E and write a paragraph to summarise the differences in appearance.
22 What will be the eventual outcome of the occluded stage?

Many depressions in midlatitudes run out of energy as they move further inland. This is certainly the case in Europe. On the west coast of North America, depressions are forced to pass over the mountain ranges, which causes a rapid loss of water vapour. During occlusion the temperature, pressure and humidity differences between the air types in a depression slowly disappear. This final dissipation of the frontal zone is called frontolysis. The air masses of the continental interiors often absorb the air types of the depression after their initial individual characteristics have been lost.

Figure 34 Depressions and upper air flow

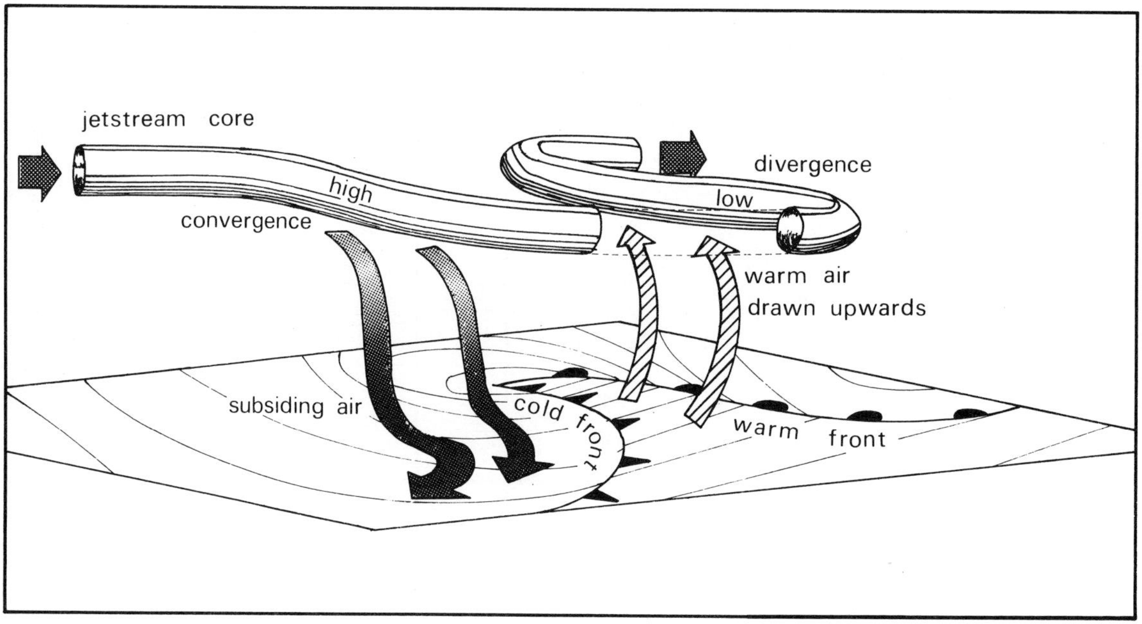

A midlatitude depression

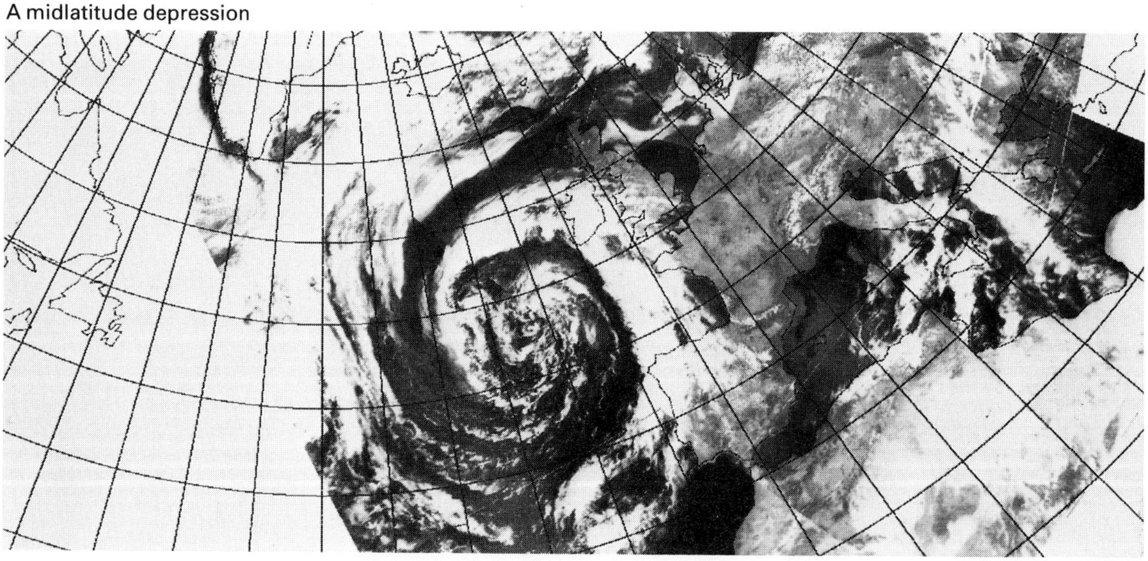

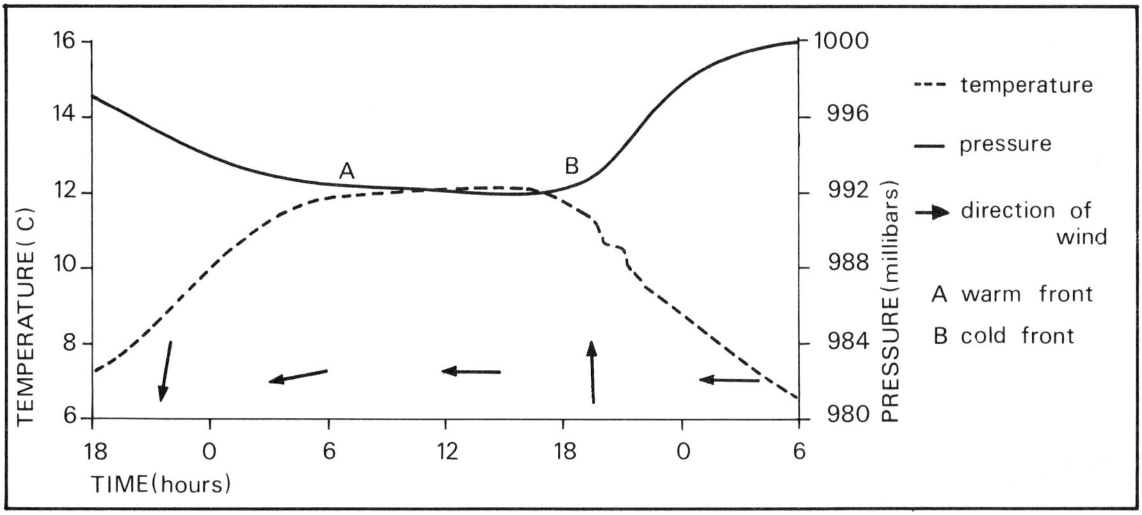

Figure 35 Changes associated with the passing of a depression

Midlatitude continental interiors

Study the details given for The Pas, Canada, 53° 58′N, 101° 06′W in Table 8.

---

23 Noting the latitude of The Pas, compare its pattern of precipitation with that of Prince Rupert (see Table 6 p.36).

24 Attempt an explanation of any differences you may have noted.

Table 8: Climate details for The Pas, Canada, 53° 58′N 101° 06′W

	J	F	M	A	M	J	J	A	S	O	N	D
°C	−22	−18	−11	0	8	14	18	16	10	4	−7	−17
mm	18	16	21	25	38	59	72	61	55	31	29	23

Interior continental areas have very large temperature ranges — even those with latitudes not far outside the tropics, e.g. Oklahoma 35° 29′N, 97° 32′W has a temperature range of 24·5°C, which is much greater than anything experienced in Britain. The major influence on the climates of continental interiors is the availability of insolation. This varies seasonally, and with increasing latitude will have a greater effect on the climate.

The Prairie region of Canada serves as an excellent example of a cool continental climate. The annual temperature ranges recorded in this area are large. Saskatoon 52° 10′N, 106° 41′W has an average annual temperature range of nearly 38°C. Moreover, the average minimum temperature for January is −24°C, while the average maximum for July is 26°C. Saskatoon has recorded an absolute minimum of −48°C and an absolute maximum of 40°C. Similar conditions exist throughout the Prairie regions of North America, and are also present in Central Russia and Siberia.

This type of climate — with a wide range of temperatures — obviously produces a distinctive annual regime. There is no gentle transition from winter to spring and then summer. In the absence of solar energy the climate is cold, the warming effect of the oceans being insignificant. Winter ends abruptly with the average last day of frost being mid to late May. Snow regularly falls in April and is often recorded in May. The average last date of snow cover in Saskatoon, for example, is May 12th, a month in which the average temperature is 11°C and the average maximum temperature is 18°C. This demonstrates the abrupt change from winter to summer experienced in the Prairie regions. Summer is short, with the first frosts being recorded in September. Winnipeg has an average of only 118 frost-free days, Regina has 107. This compares with 199 for Prince Rupert and 212 for Vancouver.

The dry Prairie air provides little moisture, except during the early summer when water is evaporated from the surface giving early summer rains. Convectional rainfall is most common with thunderstorms being a frequent occurrence. Temperatures are very high in the summer, aided by the clear skies, and commonly exceed 30°C. The likelihood of thunderstorms developing is enhanced by these high temperatures, and some of these storms produce very large hailstones which often damage the bodywork of cars.

Autumn, like spring, is very short on average, but it is dry, an important factor in the harvesting of the vast acreages of cereals. Snow appears on average in late October or early November. The date for Winnipeg is November 4th, which is very early considering its latitude. The air flow in the winter is generally from the Arctic, and is so cold that it is unable to hold large amounts of moisture — hence the low winter precipitation totals.

Winter is long and cold. Average temperatures are about −17°C since the sun is low in the sky and there is no other source of heat. It is so cold that it is inadvisable to leave a car engine switched off for even the shortest of periods; indeed most cars have heaters in their engines! Severe storms or blizzards occur during the winter. These often have winds in excess of 40km.p.h., with temperatures of less than −12°C, visibility of less than 1km and sometimes lasting for more than 6 hours.

Lapse rates, stability and instability — a case study

You will note from Chapter 1 (Figure 3) that in the troposphere, temperature decreases with altitude.

---

1 What is the rate of temperature change with altitude in the troposphere?

This general change in temperature is called the **Environmental Lapse Rate** (ELR). If you were to make an ascent in a balloon, you would experience this temperature decrease. The ELR varies over space and time, and therefore may not be viewed as being constant. When air is forced to rise, either over hills or mountains or in a depression or tropical cyclone, the air cools, and the water vapour in the air condenses. This cooling is not the product of heat loss to surrounding air, rather it is due to the ability of the air to expand as it moves to an altitude where pressure is lower. The air uses up its own energy to expand, and therefore cools. The relationship between temperature (T), pressure (P) and volume (V) is constant and may be stated:

$$\frac{P_1 \times V_1}{T_1} = \frac{P_2 \times V_2}{T_2}$$

So as pressure decreases, temperature also decreases while volume increases. If the air were to return to its original volume it would also return to its original temperature. Such temperature changes, where no heat is added or lost, are called **adiabatic**, and the rate of cooling or heating is called the adiabatic lapse rate. There are two adiabatic lapse rates: one for dry air — **Dry Adiabatic Lapse Rate** (DALR) and one for moist air — **Saturated Adiabatic Lapse Rate** (SALR). Examine Figure 36.

Figure 36 Lapse rates

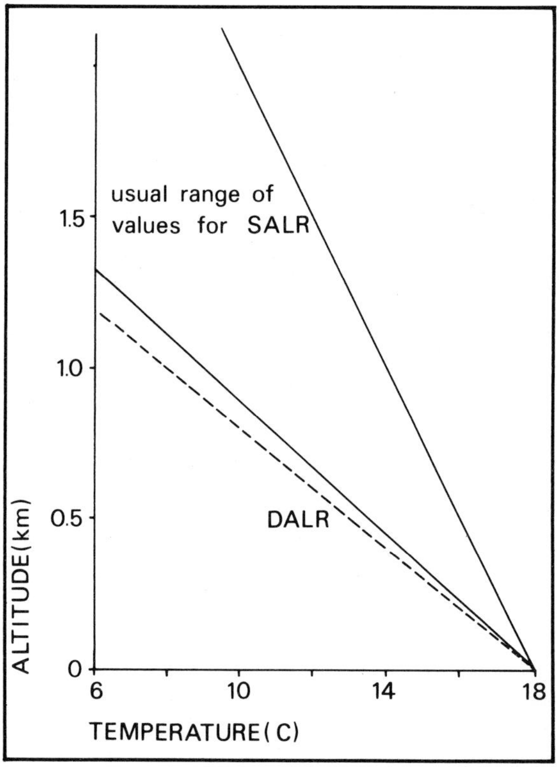

---

2 From the diagram calculate typical values for the DALR and SALR.

Under many circumstances the rising air will cool at the DALR until it reaches its dew-point — the temperature at which water vapour condenses. If at its dew-point the rising air is still warmer than the surrounding air, it will continue rising. However, as water vapour condenses, it releases latent heat which warms the air, and so the rate at which the saturated air cools (SALR) is less than the rate at which the dry air cools (DALR).

3 What factor causes the SALR to be variable?
4 Under what circumstances will the SALR be similar to the DALR?

Stability and instability

The ELR, SALR and DALR are important because they indicate the stability or otherwise of atmospheric conditions. Stability exists where air will neither ascend nor descend. Instability exists where the rising air continues to be warmer and therefore less dense than the surrounding air. Stable conditions produce only thin layers of clouds, if any, whilst unstable conditions produce very deep layers of clouds such as cumulonimbus. Examine Figure 37.

5 explain why conditions of stability are shown.

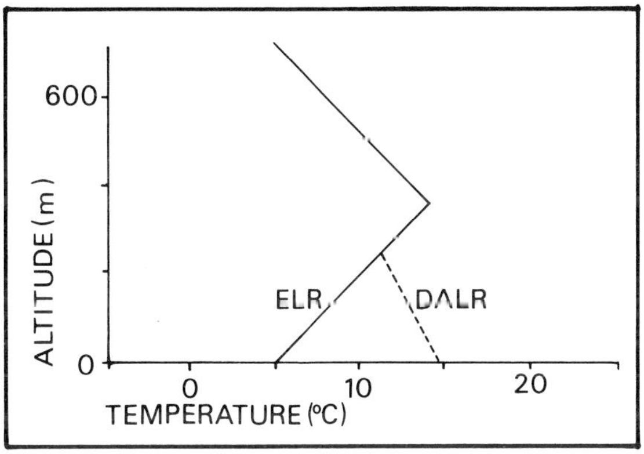

Figure 37 A temperature inversion

Examine Figure 38.

6 (i) At what altitude and temperature does the rising air become warmer than the surrounding air?
(ii) Suggest how the air was forced to rise to that altitude in the first place.
(iii) Calculate the minimum depth of the clouds.

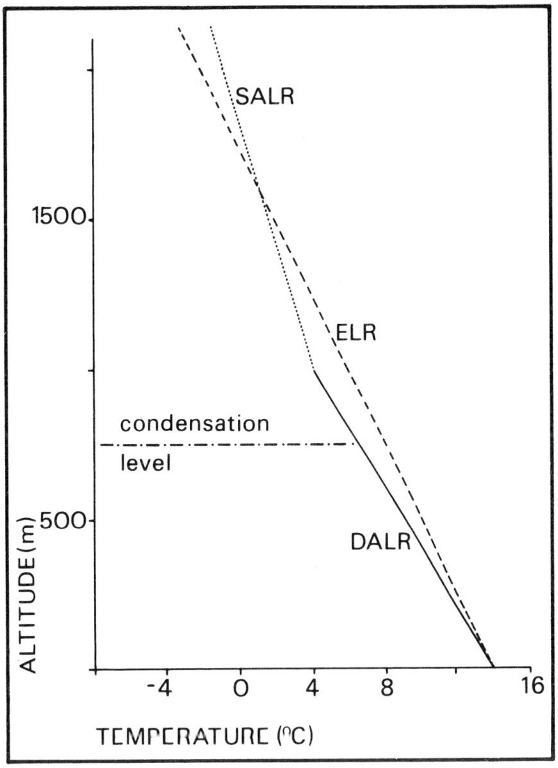

Figure 38 Conditional instability

In your answer to Question 6 you have described **conditional instability**, a situation brought about by the uplift of stable air, usually as it passes over a mountain range.

7 What happens at the point where the SALR line crosses the ELR line?

An understanding of lapse rates and the related stability and instability is the key to understanding cloud form and the likelihood of precipitation.

6 Climates of High Latitudes

High latitudes generally experience very cold climates. With the exception of locations on western coasts, average temperatures for the year are well below zero. Bergen, Norway at 60°N has an average temperature of 8°C, because of its west coast location, while Fort Smith, Canada also at 60°N has an average temperature of −3°C. Verkhoyansk, Siberia, at 67° 58′N has an average temperature of −17°C. It could be assumed that temperatures in high latitudes are always low; but this is not necessarily the case because although low temperatures are the norm, high temperatures are not uncommon, for example 37°C at Verkhoyansk and 34°C at Aklavik, Canada, 68° 18′N.

The high latitudes are dominated by the ice caps, either frozen ocean in the Arctic, or ice sheet on land in Antarctica.

---

1 What effect will the presence of such extensive cold surfaces have on the atmospheric pressure?

2 With the help of an atlas, account for the pattern of prevailing winds in high latitudes.

Midlatitude depressions may invade high latitudes, but they are unable to penetrate to any great distance, and so these latitudes tend to be quite dry.

---

3 What other feature of the air in high latitudes should ensure a lack of precipitation?

There is little marked variation in the climates of high latitudes, so we shall concern ourselves with only two types of climate: boreal and tundra.

The boreal climate

This type of climate is synonymous with the vast forests which extend from Alaska to Newfoundland and Norway to Siberia. This is a forest made up of a few hardy species which are able to survive the long cold winters and yet make the most of the short growing season.
Examine Figure 39.

---

4 Calculate from the graph the approximate temperature range and annual total precipitation.

5 Describe the annual pattern of the climate.

6 Suggest possible reasons for the summer months experiencing greater precipitation than the winter months.

Regions experiencing this climate have the largest temperature ranges in the world. Verkhoyansk, Siberia, 67° 50′N, despite its low average temperature, has an average of 13°C in July. Moreover, it has an absolute maximum of 37°C and an absolute minimum of −67°C, undoubtedly a very harsh climate and similar conditions are experienced in few other locations. In Verkhoyansk, distance from the oceans and high latitude combine to give this abnormally large temperature range.

Arctic climates

These are climates of particularly high latitudes, where there are long periods of darkness during the winter. Understandably few such places are inhabited, and then only by miners and oil company workers. The summer season is short, and temperatures rarely exceed 15°C, while average temperatures may remain below 6°C.

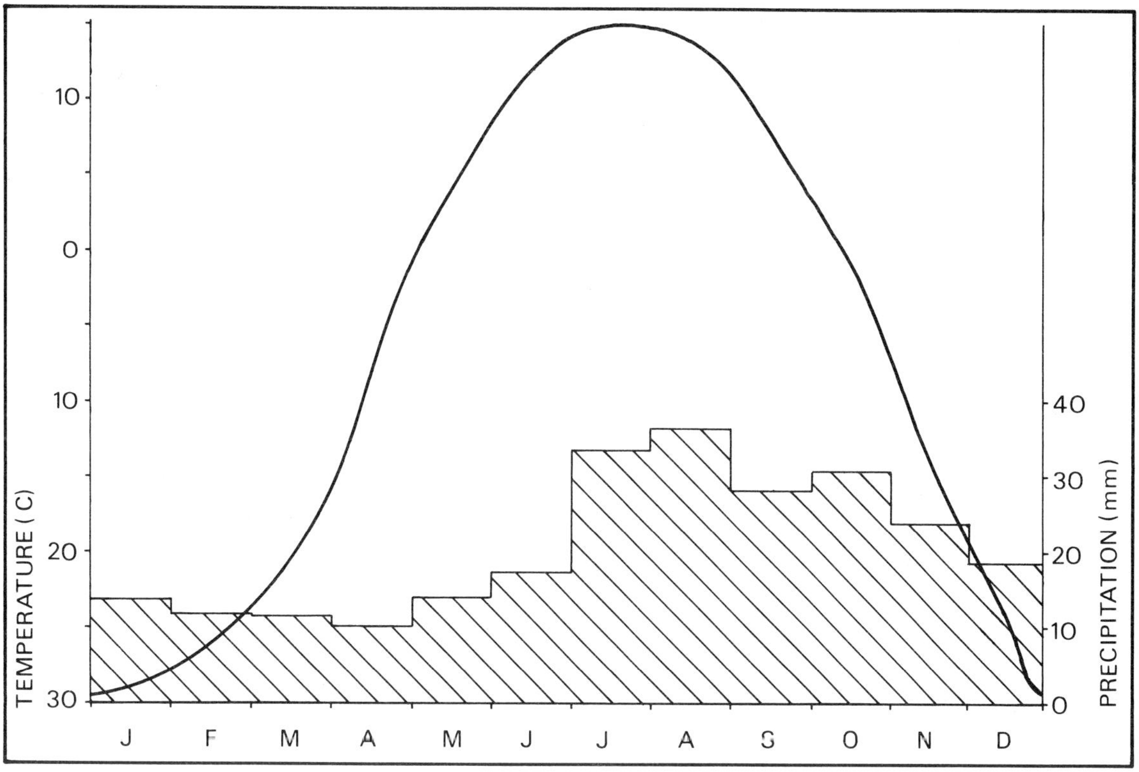

Figure 39 The climate of Yellowknife, Canada

Under such conditions plant life is very difficult, and trees absent. Plants which are able to survive under such conditions are able to operate at lower than normal temperatures. Precipitation totals are low as the very cold air is unable to hold much water vapour. The milder temperatures of the summer season do allow the air to hold margin ally more water vapour than in winter, but precipitation totals are generally below 250mm — making these areas cold deserts. Patches of snow and ice persist throughout the summer season.

7 Given the situation you have described in your answer to Question 1, draw a sketch map to show the movement of air in the winter over North America and Asia.

Examine the details given in Table 9.

8 Plot a graph of monthly temperature against precipitation (as for Figure 25 p.27).

9 Calculate the temperature range and precipitation total and make a comparison with Yellowknife (see Figure 39).

10 Referring to Figure 7 (p.12), calculate for Resolute the approximate duration of the periods with twenty-four hours daylight and twenty-four hours darkness.

Table 9: Climate details for Resolute, Canada, 74° 34′N 94° 59′W

	J	F	M	A	M	J	J	A	S	O	N	D
°C	−33	−33	−31	−23	−11	0	4	3	−5	−15	−24	−29
mm	3	3	3	6	9	12	26	30	18	15	6	5

Resolute records, on average, 321 days in the year with sub-zero temperatures. The mean date for the last snowfall of winter is June 29th, and for the first snow of the next winter September 12th. The frost-free period is even shorter, with the last frost of winter being July 10th, and the first frost of the next winter July 20th. The highest recorded temperature in Resolute is 18°C and the lowest is −52°C. To live in Resolute, you must indeed be resolute! Such climates are not supportive of complex plant life.

11 What are the main constraints on plant growth.

Survival — a case study

Windchill

At very low temperatures the velocity of the wind may have a pronounced effect on experienced temperatures. Warm days in temperate latitudes may seem chilly when there is a strong wind blowing, but in higher latitudes the strong winds coupled with low temperatures can have serious consequences for humans working and living in such environments. In particular, an organism which produces warmth as part of its life processes, has difficulty in producing heat faster than the wind removes it from the surface of its body. Very small creatures have a large surface area compared with the volume of their bodies, and are unable to produce sufficient heat in order to survive in the Arctic.

Examine Figure 40.

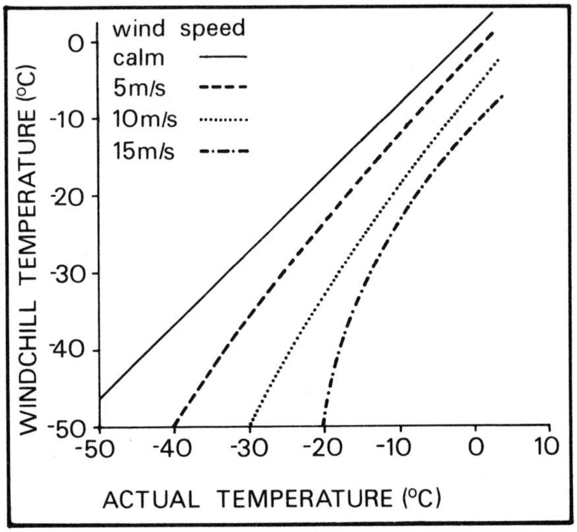

Figure 40 Windchill — the effect of wind on temperatures

1 When the wind speed is 10m/s, and the actual temperature is −20°C, what is the windchill equivalent temperature?

2 When the actual air temperature is 0°C and it is calm, why is the windchill equivalent temperature 1°C?

3 What implications does the windchill factor have for the inhabitants of the cities of the Prairies?

Figure 41 Temperature, humidity and human comfort

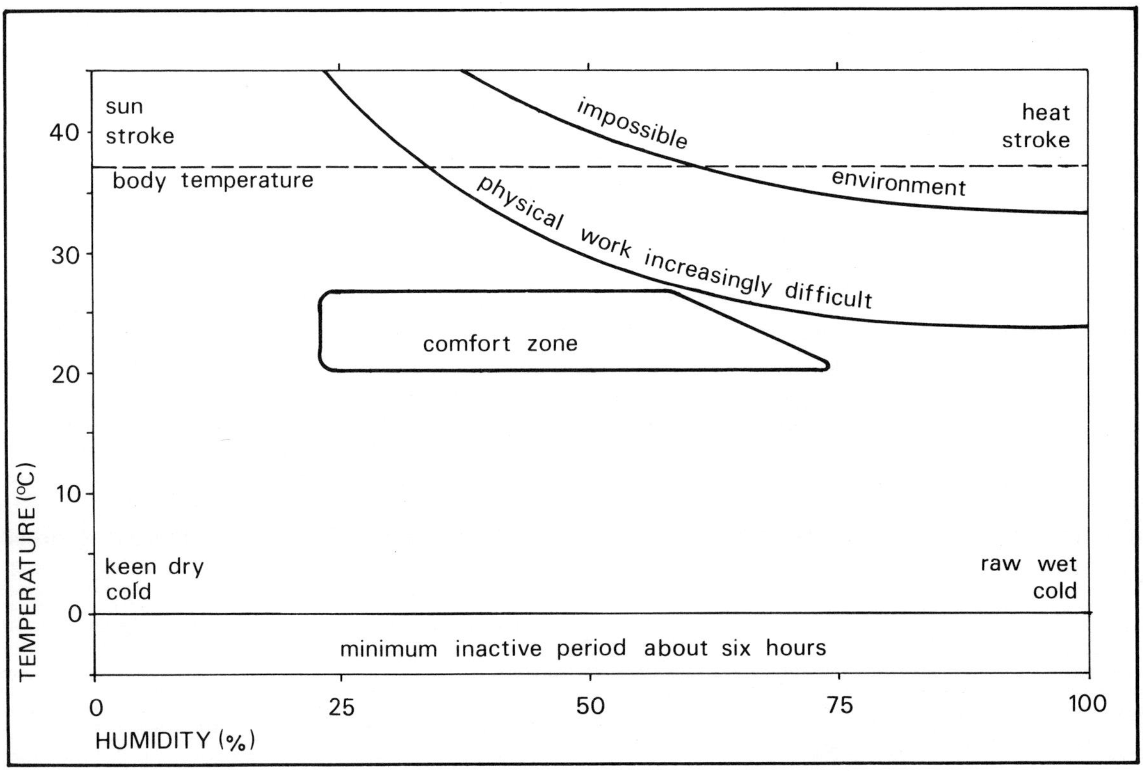

Humidity

In addition to the strength of the wind, the amount of moisture in the air is also an important factor in controlling how comfortable we feel. Under high temperatures most organisms have a cooling system which comes into operation. In humans this involves the removal of perspiration from the surface of the body, by air, taking with it excess heat. If, however, the air is already holding a lot of water vapour, i.e. its humidity is high, then the ability of the air to remove water in the form of vapour from the body will be reduced, and sweating will have little effect.
Examine Figure 41.

---

4 What is the lowest temperature at which discomfort is experienced?
5 Summarise the information contained in Figures 39 and 40, and relate this to the climates contained in this book.

Further exercises

This book is brief, and has concentrated mainly on the means by which climates are produced, rather than on the climates themselves. There are many areas which have not been touched upon, and many relationships with other areas of study, which have been hinted at but not explored.

This concluding section aims to pose questions which will lead you to further independent research into climatology and related areas.

---

1 Find out the meanings of, and make notes on, the following terms:
   water budgets — Bergeron-Findiesen Theory
   tornado — collision theory
   microclimate — interstadial
   Chinook — anabatic winds
2 The relationship between jetstreams and depressions is not as simple as suggested earlier in this book. Try to find out the precise nature of the jetstream-depression relationship.
3 In Chapter Three it was mentioned that the jetstream was an important factor in the production of the Asian Monsoon.
   (i) Describe the cycle of the monsoon in India.
   (ii) Assess the significance of the jetstream in the production of the monsoon circulation.
4 In many areas of the tropics, vast areas of rainforest are being felled indiscriminately. Assess the likelihood of this action being the cause of climatic change on both local and global scales.
5 You have probably seen and heard in the media the effects of desertification — the spreading of deserts. Is changing climate the major factor responsible for this, or are other, more important factors involved?
6 What is palynology? How has it helped in ascertaining the pattern of climatic change in Britain?
7 Human beings have the ability to alter climates either deliberately or as a result of other activities. Make notes under the following headings:
   (i) urban climates (ii) cloud seeding
   (iii) the greenhouse effect.
8 Make lists of extreme climatic events for Britain and the rest of the world. For one event from each list, describe and explain the factors by which it was produced.
9 Outline the nature of the relationship between climate, soil and vegetation in one named climatic region.
10 Make a list of landforms which may be attributed to a particular climatic regime.
11 List the factors which are of significance in the production of soils. How important is the role of climate in the development of soils?
12 Most large cities have a meteorological office. If possible, arrange a visit to your local meteorological office, and attempt to find out about the following:
   (i) data collection
   (ii) forecast accuracy
   (iii) funding.

Index